Norms in Technology

Philosophy of Engineering and Technology

VOLUME 9

For further volumes:
http://www.springer.com/series/8657

Marc J. de Vries • Sven Ove Hansson
Anthonie W.M. Meijers
Editors

Norms in Technology

Springer

Editors
Marc J. de Vries
Eindhoven University of Technology
Eindhoven, the Netherlands

Sven Ove Hansson
Royal Institute of Technology
Stockholm, Sweden

Anthonie W.M. Meijers
Eindhoven University of Technology
Eindhoven, the Netherlands

ISSN 1879-7202 ISSN 1879-7210 (electronic)
ISBN 978-94-007-5242-9 ISBN 978-94-007-5243-6 (eBook)
DOI 10.1007/978-94-007-5243-6
Springer Dordrecht Heidelberg New York London

Library of Congress Control Number: 2012954386

Printed on acid-free paper

Springer is part of Springer Science+Business Media (www.springer.com)

Preface

The book is the result of a joint effort by two groups that are involved in the philosophy of technology. At Eindhoven University of Technology, the nature of technological knowledge is a research focus for the philosophy and ethics group in the Faculty of Industrial Engineering and Innovation Sciences. At the Royal Institute of Technology in Stockholm (KTH), risks and technology is a research focus for the philosophy group in the Division of Philosophy. This book shows that there is a relation between those two focal points. Knowledge of risks in technology is one of the most important examples of normativity in technological knowledge. In the Norms in Knowledge project, which was carried out in Eindhoven in 2003–2009, various aspects of this normativity were investigated. Risks, however, did not feature strongly in this programme. That is why cooperation with the KTH group was quite beneficial. The Eindhoven programme provided a broad perspective for reflecting on normativity in technological knowledge, and the Stockholm programme provided reflection on the concept of risk as a prominent example of normativity in technological knowledge. This book is the outcome of that cooperation. We hope that it will contribute to the philosophy of technology and in particular to the reflection on the nature of technological knowledge and risks.

We want to thank all authors for their contributions. We also want to thank the Netherlands Organisation for Scientific Research, NWO, for funding the Norms in Knowledge project (project nr. 360–2011), of which this book is the final publication.

Eindhoven/Stockholm

Marc J. de Vries
Sven Ove Hansson
Anthonie W.M. Meijers

Contents

1 **Introduction** 1
Sven Ove Hansson, Anthonie W.M. Meijers, and Marc J. de Vries

Part I Normativity in Technological Knowledge and Action

2 **Extending the Scope of the Theory of Knowledge** 15
Anthonie W.M. Meijers and Peter A. Kroes

3 **Rules, Plans and the Normativity of Technological Knowledge** 35
Wybo N. Houkes

4 **Beliefs, Acceptances and Technological Knowledge** 55
Marc J. de Vries and Anthonie W.M. Meijers

5 **Policy Objectives and the Functions of Transport Systems** 67
Holger Rosencrantz

6 **Rational Goals in Engineering Design: The Venice Dams** 83
Karin Edvardsson Björnberg

Part II Normativity and Artefact Norms

7 **Valuation of Artefacts and the Normativity of Technology** 103
Sven Ove Hansson

8 **Artefactual Norms** 119
Krist Vaesen

9 **Instrumental Artefact Functions and Normativity** 137
Jesse Hughes

10 **The Goodness and Kindhood of Artefacts** 155
Maarten Franssen

Part III Normativity and Technological Risks

11 The Non-reductivity of Normativity in Risks 173
Niklas Möller

12 Risk and Degrees of Rightness 193
Martin Peterson and Nicolas Espinoza

13 Naturalness, Artefacts, and Value 207
Per Sandin

14 Trust in Technological Systems 223
Philip J. Nickel

Index 239

Chapter 1
Introduction

Sven Ove Hansson, Anthonie W.M. Meijers, and Marc J. de Vries

In order to be successful, engineers have to master and to be able to apply a wide range of descriptive knowledge, such as knowledge about the properties of materials and the workings of different kinds of constructions. That type of knowledge is similar to scientific knowledge, and its nature has been studied extensively in the philosophy of science. But descriptive knowledge is not enough. In addition, norms play important roles in engineering and technology, and engineers need to have knowledge of those too. It is the purpose of this book to clarify what these roles are and to investigate to what extent they have consequences for the nature of technological knowledge. In this introduction, we will confine ourselves to first describing the various types of norms in technology. We will then show how norms are related to central concepts in engineering, which is illustrated by a discussion of the notion of risk. At the end of this introduction, we will present the outline and architecture of the book.

1 The Many Relations Between Norms and Technology

One of the well-known examples of how technology and normativity are related is Langdon Winner's description of the social effects of the Long Island bridges over the New York parkways that were built in the 1930s.[1] Winner tells us that the bridges

[1] Langdon Winner (1980) 'Do Artefacts Have Politics?' *Daedalus* 109:121–136.

S.O. Hansson
Royal Institute of Technology, Stockholm, Sweden
e-mail: soh@kth.se

A.W.M. Meijers • M.J. de Vries (✉)
Eindhoven University of Technology, Eindhoven, the Netherlands
e-mail: a.w.m.meijers@tue.nl; m.j.d.vries@tue.nl

M.J. de Vries et al. (eds.), *Norms in Technology*, Philosophy of Engineering and Technology 9, DOI 10.1007/978-94-007-5243-6_1,

were so low that buses could not pass them, thus making it more difficult for poor black people travelling with buses to access Jones beach, a favourite public park for the white middle class. According to Winner, these bridges contributed to social segregation. They were not simply neutral means: they had a built-in politics. We can leave aside the contested issue of the veracity of this example.[2] The point that he made is certainly valid: technical systems can change the human condition in important ways by having built-in normativity. This is one important sense in which technology can be said to be normative.

New technologies sometimes make it necessary to reconsider concepts that are central in ethical deliberations, such as the concepts of health, death or community. Take, for example, the notion of death. It is notoriously hard to distinguish life from death and to mark in time the exact moment of death. Previously, death was defined in terms of cardiac arrest. But this definition became inadequate with the development of methods of cardiopulmonary resuscitation (such as defibrillation). It was also criticized because a person with permanent loss of consciousness and basic brain functions was not counted as dead. This had practical consequences for the availability of organs for life-saving transplantations. Nowadays death is legally defined in most countries as brain death, being the state in which there is an irreversible end of electrical activity in the whole brain. A flat EEG is an indication of this state, though other clinical indicators are used as well. Some have argued that this definition of brain death is too strict since only part of the brain, the neocortex, is responsible for a person's consciousness. They favour an approach in which death is defined as the cessation of electrical activity in the neocortex, making the definition even more technology dependent.

Technology can also be morally relevant by creating new options for actions. An example is prenatal screening. Its introduction gave rise to moral problems for pregnant women that did not exist before this technology was developed: should they or should they not perform prenatal tests, and what consequences should follow from the outcomes of these tests?

There are different types of norms, such as moral, legal, evaluative and instrumental norms. In engineering, moral norms are often expressed in codes of conduct and other documents that describe the responsibilities of engineers. In some areas of engineering, legal norms specify how technological systems should be designed and used. The automobile industry has to follow legal design criteria for safety and exhaust gas emissions, radio device industries have to follow regulations on the use of frequency ranges, and the pharmaceutical industry is subject to extensive regulation on the chemical and physiological properties of its products. However, most norms in engineering are neither moral nor legal. Instead, they are instrumental norms, specifying what means we should use if we wish to achieve certain ends.

---

[2] Woolgar, Steve and Cooper, Geoff (1999). 'Do artefacts have ambivalence? Moses' bridges, Winner's bridges and other urban legends in S&TS'. *Social Studies of Science* 29: 433–449.

2 Two Types of Instrumental Norms

Practical usefulness is an undisputed key value in engineering and technology. Artefacts such as computers, buildings, cars or mobile phones are designed to serve as means for practical purposes. Their quality as means is expressed in terms of instrumental norms. Two types of instrumental norms are particularly important in engineering, namely, *efficacy* norms that refer to how well artefacts realize their ends and *efficiency* norms that refer to how much input is needed to do this.

Since a technological object may realize more than one end, it is often subject to several efficacy norms. An artefact may be reliable, user-friendly, robust, cost-effective, durable, producible, maintainable, recyclable, acceptable, marketable and so on. Some of these norms refer to the purpose that the artefact is meant to serve, others to more broadly understood aims and yet others to its lifecycle and its effects on the natural and social environment. These norms all allow for degrees to which the artefact satisfies them.

Efficacy norms are important in the design of artefacts. They are part of the design specifications, which contain information about how well the artefact to be designed should perform in these various respects. Such specifications can be functional (a light bulb should produce 1,200 lm during 10,000 h of operation) or nonfunctional (the light bulb should be recyclable). Design specifications often contain conflicting requirements that give rise to trade-offs during the design process.

Efficiency is a measure of how well a certain job is done in terms of its use of resources, be it energy, materials, space, time, money, human capacities and so on.[3] In a more technical sense, efficiency is the mathematical ratio between the output and the corresponding input of a certain machine or system. Depending on the output referred to, we can, for example, speak of thermal efficiency, mechanical efficiency or electrical efficiency. There may be theoretical limits to the proportion between output and input. In the case of energy conversion, this limit follows from the second law of thermodynamics. The efficiency of an artefact or a technological system will be judged relative to these theoretical boundaries.

Efficiency tends to appeal to engineers in two ways. First, it is a measure of how well they are able to optimize the processes in the technical systems that they have designed and how close they reach theoretical limits. It is thus a measure of excellence in the profession. Engineers aim at making things better: smaller, lighter, faster, cheaper, easier to use and produce, more versatile or more energy efficient. Secondly, efficiency also has a moral appeal to engineers since it is a measure of how well the scarce resources of our planet are used. High efficiency is the opposite of wastefulness. Many of our resources are finite and will not be available to future generations if our technologies do not use them efficiently. The moral norm of sustainability or justice between generations can in part be operationalized in terms of norms of efficiency.

[3] See for an extensive discussion: Jennifer Alexander (2009) 'The Concept of Efficiency: An Historical Analysis', in A.W.M. Meijers (ed.), *Handbook Philosophy of Technology and Engineering Science*, Elsevier 2009, 1007–1030.

3 Norms, Risk and Safety

Norms are sometimes intrinsically related to central concepts in engineering. The notion of risk is a key example. We will discuss it below in more detail in order to show that there are notions in engineering that cannot be classified as purely descriptive or normative. They are to a certain extent hybrid in nature in that they have *both* descriptive and normative dimensions.[4]

Considerations of risks have been part of technology throughout its history. Technologies have been used to avert risks, but they have also given rise to new risks. In the late 1960s a new field was institutionalized: the discipline of risk analysis, with professional societies, research institutes and journals of its own.

3.1 The Illusion of Nonnormative Risk Assessment

Risk analysis arose in part as a reaction against what some described as 'emotional' and 'irrational' attitudes among the public. Risk analysts attempted to describe risks objectively, and many of them promoted the view that risk decisions should be driven by rationality alone, not by emotions or by moral or other noninstrumental norms. Based on probability calculus, a method was devised to measure risk (or its severity, the distinction was seldom made) in terms of the statistical expectation values of the outcome. The use of the term 'risk' to denote such expectation values was introduced into mainstream risk research through the influential Reactor Safety Study (WASH-1400, the Rasmussen report).[5] In probabilistic risk analysis, outcomes are weighed by their probabilities. If 1,000 workers perform a task that imposes the risk of death of 0.2% for each individual, then the expected number of fatalities from this work is $0.2\% \times 1{,}000 = 2$. Expectation values have the important property of being additive. Suppose that a certain operation is associated with a 1% probability of an accident that will kill five persons and also with a 2% probability of another type of accident that will kill one person. Then the total expectation value is $1\% \times 5 + 2\% \times 1 = 0.07$ deaths. In similar fashion, the expected number of deaths from a nuclear power plant is equal to the sum of the expectation values of each of the various types of accidents that can occur in the plant.

In the technical literature, such expectation values are often referred to as 'objective' risk. In risk studies the 'objective risk' is contrasted with the subjects' rankings of risk factors that are said to express 'subjective' or 'perceived' risk. According to the view that identifies expectation values with 'objective risk', we

---

[4] There is a clear analogy here with discussions in meta-ethics about thick and thin concepts, thick concepts being both descriptive and evaluative.

[5] Rechard, R. P. (1999) 'Historical Relationship Between Performance Assessment for Radioactive Waste Disposal and Other Types of Risk Assessment', *Risk Analysis* 19(5):763–807, p. 776.

can characterize a risk completely in terms of a certain type of objective facts about the physical world, namely, facts about possible outcomes and their probabilities. It would seem to follow that norms have no role in the analysis and management of risks; there is an objective analysis that serves us much better. However, the idea of a norm-free analysis of risks is illusory.[6] Normative issues necessarily enter the risk discourse in several ways, in particular the following three:

1. The undesirability (or acceptability) of risks
2. Prioritization among incomparable risks
3. Probability weighing

Let us consider each of them in turn.

3.2 The Undesirability of Risks

First and most fundamentally, risk implies undesirability.[7] When we say that there is a risk, we imply that something undesired may happen. We talk about the 'risk' of losing a bet but the 'chance' of winning it. Similarly, when talking about the risks associated with a surgical procedure, we refer to what may go wrong. The expected positive effects of the surgery are not called 'risks'. What is a risk for one person may not be a risk for another, simply because the risk refers to an event that the first but not the second person wishes to avoid.

Although this connection with undesirability is a simple fact about risks, it is often forgotten, largely because most of the risks we discuss concern uncontroversial values. We all agree that deaths due to road traffic accidents are undesirable, and it would be quite unnecessary to remind each other of this when we talk about death risks in road traffic. But in a more precise discussion, it is nevertheless important not to confuse an uncontroversial value statement with a value-free one.

3.3 Prioritization Among Incomparable Risks

Secondly, comparisons of risks tend to involve widely different types of undesirable effects. Surprisingly often, risk analysts have avoided this difficulty by just comparing the number of deaths that different technologies give rise to. But this is of course much

[6] Lee, T. R. 1981. The Public's Perception of Risk and the Question of Irrationality. *Proceedings of the Royal Society of London. Series A, Mathematical and Physical Sciences* 376(1764):5–16. Hansson, Sven Ove (1993) 'The false promises of risk analysis', *Ratio* 6:16–26. Boholm, Åsa (1998) Comparative studies of risk perception: a review of twenty years of research. *Journal of Risk Research* 1:135–163.

[7] Hansson, Sven Ove (2010) 'Risk – Objective or Subjective, Facts or Values?', *Journal of Risk Research*, 13:231–238.

too simplified. Many other negative consequences than death need to be taken into account, such as life-long diseases and environmental damage.

The philosophical discussion on risk comparisons has mostly been concerned with the difficulties involved in assigning an economic value to human lives. From an economic perspective it should be clear that human lives do not have a monetary price in the common sense of the word. Prices arise on markets, and in civilized societies there is no market for human lives. When a cost-benefit analyst assigns a monetary value to the loss of a human life (or to the risk of such a loss), then this value does not have the implications that a price has on a market; in particular it does not imply that anyone can buy or sell a person, or the right to kill her, at the price in question. Instead of being prices in the ordinary sense, such values indicate what society can afford to pay for measures that save a human life, given all the other social priorities that have to be taken into account, such as schools, crime prevention, healthcare and the rest. Therefore, the 'values of life' applied in a society reflect the resourcefulness of that society at least as much as they reflect how much it values human lives. A poor country cannot pay as much to save a human life as a rich country, but this does not mean that it values the lives of its citizens less.[8]

In the analysis of technological risks, there are usually several types of negative outcomes that need to be taken into account, such as deaths, diseases, environmental damage and loss of cultural heritage. Unfortunately there is no generally accepted way to compare these different types of outcomes. There is no definite answer to the question how many cases of juvenile diabetes correspond to one death, or what amount of human suffering or death corresponds to the extinction of an antelope species. The fundamental problem here has nothing to do with money. It consists in the predicament that for practical purposes we need to comparatively evaluate entities that we have good reasons to consider being incomparable. The idea that this could be done in an objective fashion, free from values and norms, is chimerical.

3.4 Probability Weighing

Thirdly, normativity enters the risk decision process through our choices on how to use probabilities in risk decisions. As mentioned above, it is often taken for granted in risk analysis that expectation values constitute the only correct way to use probability information. It is, for instance, assumed that a risk of 1 in 10,000 that 20,000 people will die is equally serious as a risk of 1 in 2 that 4 people will die. However, this approach is normatively problematic in at least two ways. First, it cannot be taken for granted that the moral impact of a potential outcome is proportionate to its

[8] Hansson, Sven Ove (2007) 'Philosophical Problems in Cost-Benefit Analysis', *Economics and Philosophy* 23:163–183.

probability. Some discussants claim that the avoidance of large disasters, such as a nuclear accident costing thousands of human lives, should be assigned a higher priority than what is warranted by the statistically expected number of deaths.[9] Whether or not such risk-averse or cautious decision-making is preferable is a normative issue that cannot be reduced to nonnormative issues.

Furthermore, expected utility calculations can be challenged for not respecting individual rights. For example, suppose that in an acute situation we have a choice between two ways to repair a serious gas leakage in the machine room of a chemical factory. One of the options is to send in the repairman immediately. There is only one person at hand who is competent to do the job. He will run a risk of 0.9 to die due to an explosion of the gas immediately after he has performed the necessary technical operations. The other option is to immediately let out gas into the environment. In that case, the repairman will run no particular risk, but each of the 10,000 persons in the vicinity of the plant runs a risk of 0.001 to be killed by the toxic effects of the gas. In this case, in order to minimize 'risk' (the expected number of deaths), we would have to send in the repairman to die. However, a rational decision-maker may refrain from minimizing expected damage due to respect for the repairman's moral rights.

3.5 Safety Norms in Engineering Practice

In conclusion, normativity enters our decisions about technological risks in several ways. It should therefore be no surprise that norms have a central role in almost all engineering approaches to risk. Safety engineering is an under-theorized activity, and it still remains to investigate how common systems of safety norms are constructed and how they relate to more general normative principles for risk abatement.[10] Here, as a preliminary input to that research, we will emphasize three common features of engineering safety norms.

The first feature is that safety norms assume that things can go wrong. This includes, in particular, recognition that humans sometimes make strange mistakes. Therefore, even if we have done everything we can to prevent an explosion in our storage of inflammable chemicals, we should keep it at a distance from places where people work regularly. Even if you do not intend to touch the energized parts in the interior of a machine, you should turn off the electricity before removing the cover.

---

[9] O'Riordan, T. and J. Cameron, eds. (1994) *Interpreting the precautionary principle*. London: Earthscan. O'Riordan, T., J. Cameron, and A. Jordan (eds.) (2001) *Reinterpreting the precautionary principle*. London: Cameron May. Burgos, R. and O. Defeo (2004) 'Long-term population structure, mortality and modeling of a tropical multi-fleet fishery: the red grouper Epinephelus morio of the Campeche Bank, Gulf of Mexico', *Fisheries Research* 66:325–335.

[10] Möller, Niklas and Sven Ove Hansson (2008) 'Principles of engineering safety: risk and uncertainty reduction', *Reliability Engineering and System Safety* 93:776–783.

The second feature is that safety norms add an extra margin to the requirements on technological artefacts and processes. Long before written history, builders and toolmakers added extra strength to their constructions to be on the safe side. Since the latter half of the nineteenth century, such practices have become increasingly mathematized. Their most common mathematical form is that of a safety factor, by which is meant a prescribed ratio of the maximal load not inducing failure to a corresponding measure of the applied load. If we build a bridge strong enough for twice the expected maximal load, then this means that we apply a safety factor of 2.

In everyday technological practices, safety margins are often applied without the guidance of exact, predetermined numerical values. When you back up to a safe distance after lighting fireworks or hold your thumb a couple of centimetres away from the nail that you are aiming at with the hammer, you apply a safety margin. The same is true when as a driver you keep a distance to the car in front of you. (In studies of driver behaviour, this margin is measured as the headway, i.e. the time it would take to drive the distance between the cars.)

The third feature of safety norms is that they are not subject to exceptions based on economic considerations. Exceptions from a safety norm can be made for safety-based reasons, i.e. in cases where adherence to the norm would not in fact increase safety. However, such exceptions should not be made for economic reasons. The fact that a particular company cannot afford to follow a safety norm is not a reason for exemption.

From an economic point of view, it might seem more sensible to adjust the safety demands to the costs, for instance by weighing the costs of safety measures against the expected safety effects in each individual case, in order to maximize the risk reduction obtained for the total expenditure on safety. But there are psychological reasons why technological safety is better served by general norms than by optimization procedures. It is in practice much easier to uphold general norms that apply to everyone than to implement the outcome of case-by-case optimization.[11] Judging by experience, the consistent application of simple norms is an essential component in successful safety engineering.

4 The Structure of the Book

The discussion above strongly suggests that norms in technology play a role that goes beyond the traditional role of epistemic norms in the natural sciences. Key concepts in engineering are not purely descriptive but have a normative dimension as well. This calls for further investigations. This book attempts to clarify a number of issues that arise from these observations.[12]

[11] Doorn, Neelke and Sven Ove Hansson (2011) 'Should safety factors replace probabilistic design?', *Philosophy and Technology* 24:151–168.

[12] Similar observations can be made about standards in engineering, an important topic that is only briefly discussed in this volume.

In Chap. 2 Anthonie Meijers and Peter Kroes argue that the nature of technological knowledge poses questions to traditional epistemology. The 'justified true belief' account of knowledge has already been challenged for many reasons, but one that so far has not been brought forwards is that it cannot account for various types of technological knowledge, such as functional knowledge, prescriptive knowledge and design knowledge. In their view technological knowledge is a key example where the standard approach fails. In Chap. 3, Wybo Houkes discusses the existing literature on normativity in technological knowledge, as expressed in technological rules and norms. He argues that it does not answer questions such as how prescriptive knowledge relates to theoretical (descriptive) knowledge and what types of actions could be involved in prescriptive knowledge. Houkes uses the notions of plans and agents to obtain a better account in which such questions do get an answer. He then argues that the normativity in prescriptive knowledge need not be derivable from scientific laws, as Mario Bunge claimed. This suggests that prescriptive knowledge is a distinct kind of knowledge. Another argument for acknowledging this distinct nature is presented in Chap. 4 by Marc de Vries and Anthonie Meijers, who argue that at least part of technological knowledge is not of a belief-based kind, but is acceptance-based.

In the next two chapters, two case studies are presented in which this distinct nature of technological knowledge is investigated in more detail. In Chap. 5, by Holger Rosencrantz, the case of Swedish transport systems is used to make three claims: (1) goals in engineering are subject to evaluation and updating, (2) there is a trade-off between precision and flexibility in the act of goal setting and (3) different kinds of goals require different approaches to implementation. In Chap. 6, by Karin Edvardsson Björnberg, the case of the Venice dams is presented to show that technological knowledge, used to define goals in technological design processes, is not evaluated on the basis of truth but on the basis of other criteria such as feasibility, precision, evaluability, approachability and consistency. These criteria are rational, and the author argues that non-derivability from empirical knowledge does not mean that the evaluation of such knowledge is a matter of irrational preferences.

Chapters 7 through 10 focus on artefact-related knowledge. In Chap. 7, Sven Ove Hansson shows that people use normative knowledge claims about artefacts continuously ('I know that this is a good screwdriver') and that there are alternative ways to classify such claims. This confirms, he argues, the idea expressed earlier in this introductory chapter, that normativity is intrinsically related to technology. This is also shown by Krist Vaesen in Chap. 8. He aims to classify artefactual norms and distinguishes between norms based on compatibility (with other artefacts), on interference (with other artefacts) and on quality. These classes are mainly of interest for the users. For designers and producers, in addition, Vaesen identifies classes of norms based on marketability, manufacturability, transportability/installability and recoverability.

The next two chapters investigate how this artefact-related normativity features in the context of human actions. In Chap. 9 Jesse Hughes shows how technological knowledge functions in the context of practical reasoning, the type of reasoning that results in directives for action rather than descriptive claims about reality. In his text he particularly investigates knowledge of functions, as artefact functions are directly related to actions. In Chap. 10, Maarten Franssen, too, uses the notion of functionality to show that norms applying to artefact kinds do not necessarily apply to sub-kinds. Hence, we have to look for a more complicated logic for such norms, and Franssen draws from the notion of use plans, brought forwards previously by Houkes in Chap. 3, to develop such logic. According to this analysis, normative judgements can be objective, but need not be.

The final four chapters deal with the notion of risk as a particular and important example of a normative notion in technology about which we can have knowledge. In Chap. 11, Niklas Möller argues that the normative notion of risk cannot be reduced to nonnormative notions. This confirms what Houkes also argued for in Chap. 3, namely, that the normativity in technological knowledge is not derivable from or reducible to nonnormative knowledge content. This notion of risk is further explored by Peterson and Espinoza in Chap. 12. The authors show that for risk, normativity comes in degrees. Risky actions are not per se right or wrong in an absolute sense, but they can be more or less right and more or less wrong. An interesting aspect in their text is that they show that this has implications for general debates in normative ethics. This is another example of philosophy of technology having implications for general philosophy. In Chap. 13 Per Sandin shows that normativity also comes in degrees when we make environmental judgements for artefacts. He distinguishes between artefacts that are more or less 'natural' and then discusses whether or not the more natural artefacts should be judged more positively when it comes to environmental issues.

Finally, in Chap. 14 Philip Nickel adds a further important notion when it comes to normativity and risks, namely, trust. Nickel argues that trust in technological systems is more than a reliability statement about those systems and also that it cannot be reduced to trust in the persons who designed and made the system. The fact that it cannot be reduced to a reliability statement again confirms that the normativity we are dealing with cannot be reduced to descriptive, theoretical knowledge. Nickel identifies an attitude of entitlement to the system's performance as a necessary element in trust, in addition to a belief that the system is likely enough to perform in the desired way. Consequently, the ethics of trust is not just a matter of an ethics of belief (name in the system being likely enough to perform safely). It also entails pragmatic factors affecting what one feels entitled to from the technological system.

In summary, what these analyses of normativity in technological knowledge strongly suggest is that:

- This normativity cannot be derived from or reduced to nonnormative notions (Meijers and Kroes, Houkes, Möller, Nickel).
- It has an element of voluntariness but this does not make it irrational (de Vries and Meijers, Rosencrantz, Edvardsson Björnberg).
- The normativity can be expressed in norms that can apply to both goals (Edvardsson Björnberg) and means (Vaesen, Franssen, Hansson) and to the practical reasoning that connects the two (Hughes).

- As the particular case of risks shows, this normativity comes in degrees (Peterson and Espinoza, Sandin).

Together the contributions in this volume show that normativity is linked to technological knowledge in a variety of ways that go beyond the normativity of epistemic norms. Many of the issues explored here are of potential interest to more general philosophical discussions and deserve further investigations.

Part I
Normativity in Technological Knowledge and Action

Chapter 2
Extending the Scope of the Theory of Knowledge

Anthonie W.M. Meijers and Peter A. Kroes

1 Introduction

The invention of the light bulb is without any doubt one of the major inventions of the nineteenth century. It changed life profoundly in that it made human activities independent of the natural light circumstances on a hitherto unknown scale. The idea of electric light goes back to Humphrey Davy. He discovered that an electric arc between two poles could produce light. This phenomenon, however, was rather unpractical for home applications and could not be an alternative to existing gas light. New ideas were explored on the basis of the theoretical work on heat production in resistant conductors. An electric current could heat a conductor to such an extent that it would glow white-hot. Joseph Wilson Swan was the first to construct an electric light bulb on these physical principles. It contained a thin strip of material (filament) in an oxygen-free environment to prevent the strip from burning. An electric current heated the strip. He had, however, serious trouble maintaining the vacuum in the light bulb. Thomas Edison was able to solve this problem and at the same time to construct stable enough carbon wire that could function as filament. On October 21, 1879, he illuminated an electric lamp that glowed continuously for 40 h. That was an enormous achievement in those days.

US Patent 223,898, filed by Edison in November 1879, is about the design and manufacturing of electric lamps. The patent describes the claim Edison makes as follows:

A.W.M. Meijers (✉)
Eindhoven University of Technology, Eindhoven, the Netherlands
e-mail: a.w.m.meijers@tue.nl

P.A. Kroes
Delft University of Technology, Delft, the Netherlands
e-mail: p.a.kroes@tudelft.nl

M.J. de Vries et al. (eds.), *Norms in Technology*, Philosophy of Engineering and Technology 9, DOI 10.1007/978-94-007-5243-6_2,

I claim as my invention:

1. An electric lamp for giving light by incandescence, consisting of filament of carbon of high resistance, made as described, and secured to metallic wires, as set forth.
2. The combination of carbon filaments with a receiver made entirely of glass and conductors passing through the glass, and from which receiver the air is exhausted, for the purposes described.
3. A carbon filament or strip coiled and connected to electric conductors so that only a portion of the surface of such carbon conductors shall be exposed for radiating light, as set forth.
4. The method herein described of securing the platinum contact-wires to the carbon filament and carbonizing of the whole in a closed chamber, substantially as set forth.[1]

The method Edison mentioned here consists of the following:

If the carbon thread is liable to be distorted during carbonization it is to be coiled between a helix of copper wire. The ends of the carbon or filament are secured to the platina leading-wires by plastic carbonizable material, and the whole placed in the carbonizing-chamber. The copper, which has served to prevent distortion of the carbon thread, is afterwards eaten away by nitric acid, and the spiral soaked in water, and then dried and placed on the glass holder, and a glass bulb blown over the whole, with a leading tube for exhaustion by a mercury-pump. This tube, when a high vacuum has been reached, is hermetically sealed.[2]

Light bulbs of the type developed and produced by Edison and his contemporaries have been for a long time the dominant type of electric lighting. They represented the 'normal design' of an electric lamp. This is not to deny the large improvements made in the construction, material and production of these lamps over the years, as well as the many alternatives that were developed and that are increasingly popular. But the basic layout and the physical principles on which the standard light bulb is based remained more or less the same.

Using Edison's patent, we will develop in Sect. 2 an initial classification of the types of knowledge involved in engineering, and we will make a few of observations about the relationship between science and engineering. We will then address in Sect. 3 the question whether current epistemological theories can be used to analyse engineering types of knowledge. The answer will largely be negative. In Sect. 4 we will develop a preliminary account of the various types of engineering knowledge. This will be followed in Sect. 5 by an attempt to identify the central issues that need to be addressed in this field, among which are the question whether knowledge has to be belief based and the issue of the relation between knowing and acting. The conclusion summarizes our main findings.

[1] Quotations are taken from the patent description found at the extensive Edison website at Rutgers University: http://edison.rutgers.edu/. See also the Appendix.

[2] ibid.

2 Science and Engineering Knowledge

Edison's patent points to several *types* of knowledge involved in the invention and making of electric lamps. An initial classification from the point of view of engineering practice would include knowledge of physical and chemical phenomena; for example, when Edison uses an electric current to heat a carbon wire, he knows about Joule's law, which relates the amount of heat produced by an electric current and the resistance of an electric wire. He also knows about the physical and chemical properties and dispositions of the various materials of the lamp: carbon, cotton, glass, platinum or copper. Let us call this category *structural knowledge*. Secondly, there is *functional knowledge* of the various parts of the electric lamp and of the equipment for producing it. For example, Edison mentions in his patent burners, receivers, clamps and leading wires. These are functional characterizations, and not physical descriptions. Thirdly, the patent describes procedures for manufacturing the carbon conductors of high resistance (see above) and for securing perfect contact between the metallic conductors and the carbon conductors. We will call this *prescriptive knowledge*. Fourthly, the patent contains a design concept that describes the layout of the electric lamp and how its parts contribute to the overall aim of producing electric light. Such a design concept usually includes a drawing or a schematic representation of the main parts of the artefact (see Appendix). Design concepts are part of a larger category of *design knowledge*, which contains knowledge about how to proceed from a functional to a structural description of an artefact, or vice versa as in the case of reverse engineering. Finally, practical *know-how* is mentioned in the patent, for example, know-how to create a vacuum, or know-how to make a carbonized wire with a cotton kernel, or know-how to make an electric connection between the wire inside the vacuum tube and the outside environment.

Based on this rough classification of the types of knowledge involved in Edison's patent, two observations can be made. The first is that the idea of technology or engineering as applied natural science is very hard to defend. It would mean that of the five types of knowledge, four could somehow be derived from the first type. The claim would be that by applying knowledge of physical principles and properties to the problem at hand (finding artificial means for lighting), you are able to arrive at the design concept of Edison's electric lamp, and at the procedural knowledge, functional knowledge and know-how which are necessary for manufacturing the carbon wire to be used in this lamp. That sounds like a very implausible and hard-to-defend claim.

Many philosophers and historians of technology have made the point before that engineering is not simply applied natural science. Historical studies by Edwin Layton and others, for example, showed that attempts to make top-down applications of Maxwell's theory of electricity actually *inhibited* the development of electric motors (see Mitcham 1994, 202). Bottom-up development was needed, with some general guidance from Maxwell's equations of course. Similarly, Joule's law or Ohm's law did not simply lead to the invention of the electric lamp. Edison used knowledge of these physical principles when he improved existing designs, but the design itself cannot be said to be a direct application of these laws. There are even cases where the physical principles underlying a device were not understood at all when designing and constructing it, as for example in the case of the early steam engines. Engineers will also readily agree that technological devices are almost never completely understood in a scientific sense. Scientific theories are about idealized phenomena, whereas engineering deals with real-life cases. These cases are usually too complicated for a purely natural science approach, and rough approximations and ad hoc models are part and parcel of engineering. So not only are there *types* of knowledge in engineering that do not fit the physical knowledge type (design concepts, procedural knowledge, functional knowledge, know-how); the knowledge actually produced by natural science, though extremely important for engineering, is often not sufficient for the description and explanation of the complex physical phenomena engineers deal with.

A second and related observation concerns the division of the sciences in general. It is sometimes claimed, for example by James McAllister, that the modern sciences can be divided into two groups (see McAllister 1997). Both groups aim at describing the world, though in different ways. The groups are defined by the type of knowledge that the sciences characteristically produce. On the hand, there are sciences that produce laws of nature. Laws are statements of universal scope, usually expressed as mathematical equations, and true of idealized phenomena. Modern physics is a clear example of this so-called law-formulating type of science. On the other hand, there are sciences that produce singular statements about actual occurrences. They describe and explain particular phenomena or particular properties of individual entities at a particular moment in space and time. Low-level generalizations produced by these sciences are best conceived as conjunctions of these singular statements. Large parts of biology (botany, evolution history), the earth sciences and the engineering sciences are regarded as examples of this so-called natural-historical type of science. They aim at bringing to the fore the variety and detail of individual phenomena, as opposed to universal regularities.

What to think of the claim that a large part of the engineering sciences fits the second category? If we look at Edison's patent and the five types of knowledge involved, there are certainly empirical generalizations that fit the natural-historical science approach. For example, when Edison describes 'that even a cotton thread properly carbonized and placed in a sealed glass bulb exhausted to one-million of an atmosphere offers from 100 to 500 Ω resistance to the passage of the current, and that it is absolutely stable at very high temperatures'.[3] But there are also types of knowledge that do not fit either category of science. Design concepts, procedural knowledge and functional knowledge are difficult to classify in terms of the two categories. They are neither natural laws nor empirical generalizations. This may be partly due to the fact that the aim of engineering science is not simply to describe (parts of) the world which is the common aim of law-formulating and natural-historical types of science. Engineering is as much about changing the world as it is about describing it.

[3] See Appendix, page 2.

So engineering science appears to be different from applied physical science, whether that physical science is of the law-formulating or natural-historic type. It is rather a third type of science, which generates not only structural knowledge of the world but also functional knowledge of artefacts, prescriptive knowledge how to achieve things, design knowledge and various forms of know-how that go together with these types of knowledge. If this observation is right, the question arises: what can modern theories of knowledge offer to analyse these types of knowledge? That will be the topic of the next section.

3 Engineering Knowledge

Modern epistemology is almost exclusively concerned with beliefs, their genesis and their justification. Monographs, textbooks, collections of essays or survey articles typically are about scepticism, justified true belief (where justification may be individual or social), counterexamples of the Gettier type, contextualism (the theory that justification is context-bound), foundationalism and coherentism, internalism/externalism debate, fallibilism and reliability, perception and epistemic norms (Martin 1998; Pryor 2001; Sosa and Kim 2000). The central question is usually conceived as follows: 'What grounds do we have for believing?' (Pollock 1986) or 'Which beliefs are justified and which are not?' (Dancy 1985). Adam Morton distinguishes three central questions that the theory of knowledge tries to answer: 'What qualities should our beliefs have?' 'What qualities do our actual beliefs have?' and 'What qualities could our beliefs have?' (Morton 1997). Robert Audi in his introduction to epistemology takes belief to be the central notion and analyses the various sources for justification (Audi 1998). At the end of his book, he discusses scientific, moral and religious knowledge. The issue here is whether moral and religious knowledge is to be interpreted in terms of beliefs, and if so, what are the sources for and the nature of the justification of these beliefs? More recent textbooks, for example (Lemos 2007), also do not go beyond these questions.

None of these books, considered to be central to the current epistemological canon, develop a systematic account of types of knowledge other than those that are belief based. There is a reference to procedural knowledge by Pollock in the context of epistemic norms (Pollock 1986, 129). These norms are to be understood as internalized rules for reasoning. Design concepts, functional knowledge and know-how are not discussed. Authors such as Ryle, who made the distinction between knowing how and knowing that, or Polanyi with his analysis of skills, are almost completely ignored in these texts (Polanyi 1978). This suggests that modern epistemology has little to offer with respect to engineering knowledge.

As may be expected, things are a bit better in the philosophy and history of technology, where several authors have attempted to make a classification of the types of knowledge involved in engineering (see Mitcham 1994, Chapter 8). One of the best to date is Walter Vincenti's well-known taxonomy (see Vincenti 1990, Chapter 7). He distinguishes between six categories of engineering knowledge:

fundamental design concepts, criteria and specifications, theoretical tools, quantitative data, practical considerations and design instrumentalities. The category of fundamental design concepts contains the normal layout of the device and its operational principles (which in turn define success and failure of the device). The second category of criteria and specifications contains technical requirements that a device needs to fulfil. In order to specify them, qualitative goals for the device need to be translated in concrete quantitative goals. Often a range is defined within which a device needs to be. The third category of theoretical tools includes scientific theories, mathematical methods and theories for design calculations. Quantitative data, Vincenti's fourth category, contains both descriptive knowledge (physical constants, properties of substances, chemical reaction velocities, etc.) and prescriptive knowledge (safety factors, standards set by industry, etc.). The fifth category, practical considerations, consists of practical knowledge that is relevant to the case at hand. Experience is an important element of this category, as is sound judgement. The last category is called design instrumentalities and contains knowing how and procedural knowledge. Optimization is an important procedure in engineering, and engineers know how to do that.

Vincenti's taxonomy is set up typically from an engineer's perspective. He describes the knowledge toolbox available to engineers when they try to solve design problems. It is thus a more process-oriented taxonomy. He is not so much interested in the nature of the instruments in the engineer's knowledge toolbox as in what instrument to use for solving what kind of engineering problem when designing and making technical artefacts. His categories are mixed in terms of the earlier taxonomy. The category of quantitative data, for example, contains both structural knowledge and prescriptive knowledge. Or the category of design instrumentalities contains know-how and procedural knowledge. It is fair to say that though Vincenti's categorization was a tremendous step forwards, there is still much to be desired with respect to the analysis of the types of knowledge involved in engineering. It is thus time to make a fresh start.

4 Exploring Types of Engineering Knowledge

Let us return to Edison's patent of his incandescent lamp and the initial classification we made. Like Vincenti's, this classification is made from the point of view of engineering practice, but contrary to Vincenti's, it aims at listing epistemologically different categories. In the practice of designing, developing and making of technical artefacts at least, the following different types of knowledge can be distinguished:[4]

[4] We will concentrate here mainly on knowledge with regard to a specific type of technological artefacts, namely, physical/chemical artefacts. We expect, however, that our conclusions with regard to the various types of knowledge involved in this kind of technological artefacts may be generalized to biological artefacts and socio-technical systems. This may not be true for software (probably best conceived as an incomplete artefact).

- *Structural knowledge*. By structural knowledge, we mean knowledge of the physical/chemical properties of the artefact. A technical artefact is an object with a certain function, and this function is to a large extent realized by its physical structure. Knowledge of this physical structure, that is, of the artefact as a physical object, is of prime importance for engineering practice. A structural description of an artefact is a description of the thing in the language of physics (and chemistry). It is a description of its static properties, such as its geometry, its mass and the composition of the materials used, and its dynamic properties, such as its mechanical, thermodynamical, electromagnetic and chemical behaviours. In Edison's electric lamp patent, a description of these properties is to be found throughout the specification, for example, in phrases such as 'The current is conducted into the vacuum-bulb through platinum wires sealed into the glass' and in the drawing, which gives the geometrical layout of the lamp.
- *Functional knowledge*. Looking at Edison's lamp from a functional point of view, the central question is the following: What is the function of this object and how does this thing work? The answer is a functional description of the thing, which contains a characterization of the overall function and of the sub-functions of its parts. This description represents the thing as a technological artefact and details how it is designed, how its structure is related to its function, how it behaves as a technological artefact and how it should be used. In Edison's patent functional descriptions are to be found throughout the specification in phrases such as '*h h* [referring to the drawing] are the clamps, which serve to connect the platina wires, cemented in the carbon, with the leading-wires *x x*'.[5]
- *Prescriptive knowledge*. Technical artefacts are not simply given, but designed and made by engineers. What kind of knowledge is required to make them? Making things is an essential aspect of technology and the knowledge involved in making things an essential component of technological knowledge. It consists, among others, of procedures. A procedure is a series of prescribed actions that will lead to a desired result: the produced artefact. Procedures also play a role in the design, for example, when optimization procedures are used, and in the operation of an artefact (as in 'switch off the power before opening the case'). Prescriptive knowledge is also contained in Edison's patent; he explicitly writes: 'The invention further consists in the method of manufacturing carbon conductors of high resistance, so as to be suitable for giving light by incandescence, and in the manner of securing perfect contact between the metallic conductors or leading-wires and the carbon conductor.'
- *Design knowledge*. Just as making things is an essential aspect of technology, so designing things is. The knowledge involved in designing consists of, among others, design concepts and operational principles, as we have seen before.

[5] It is important to note that in engineering practice structural and functional descriptions are often mixed; Edison's description of his invention is a clear example of this. This does not mean, however, that purely structural and functional descriptions play no role in engineering practice; shop floor drawings with specifications of the materials to be used are an example of purely structural descriptions and functional decompositions of complex artefacts, an example of purely functional descriptions.

More generally, design knowledge is knowledge of how to realize functions by inventing and making physical/chemical structures for users. Apparently, this aspect of technical knowledge was of paramount importance to Edison; the proud opening sentence of his patent reads: 'Be it known that I, THOMAS ALVA EDISON, of Menlo Park, in the State of New Jersey, United States of America, have invented an Improvement of Electric Lamps, and in the method of manufacturing the same….'

- *Know-how*. The design, development and making of technical things require not only knowledge of facts but also know-how in the sense of skills. This is more or less self-evident if we take into consideration the prescriptive nature of much of engineering knowledge. The adequate performance of prescriptions often requires a lot of skills. Obviously, this kind of knowledge is not explicitly contained in Edison's patent, but is implicitly presupposed when he writes: 'that carbon filaments may be made by a combination of tar and lamp-black, the latter being previously ignited in a closed crucible for several hours and afterward moistened and kneaded until it assumes the consistency of thick putty. Small pieces of this material may be rolled out in the form of a wire as small as seven one-thousands of a inch in diameter and over a foot in length …'. These are prescriptions, but clearly the adequate performance of these directions requires know-how in the form of skill.

The taxonomy suggests that technological knowledge involves at least knowledge about (physical/chemical) *structures*, *functions* and *actions* and that it may be of a *descriptive* and *prescriptive* nature. With regard to actions, technological knowledge is often of a prescriptive nature: it tells how to achieve certain ends through actions, which involves *know-how*. Normativity is also involved in the evaluative claims that play an important role in engineering practice: objects may malfunction, one series of actions may be better to realize an aim than another, and one object or design may be better in fulfilling a function than another.[6]

5 Will the Justified True Belief Account Work?

Up till now, epistemology has focused almost exclusively on the justified true belief approach, which is specifically appropriate for the first type of knowledge we have distinguished: *structural knowledge*. The standard tripartite account of (propositional) knowledge holds that agent A knows that *p* if and only if:

1. *p* (i.e. *p* is true).
2. A believes *p*.
3. A's belief that *p* is justified.

[6] We will not discuss extensively the kinds of knowledge involved in the *use* of technical artefacts. At first sight, it appears that this will not add any new types of knowledge to the ones mentioned so far. Clearly, prescriptive knowledge is important in this context (cf. the user manual) and know-how.

We all know that there are problems with this account (the Gettier counterexamples), and a lot of work has been done to remedy this account by trying to identify additional criteria for knowledge. Still, the elements of belief, justification and truth are widely regarded as necessary elements for an account of knowledge, though together they might not be sufficient.[7]

This definition of knowledge applies to any kind of proposition *p*, independently of whether *p* stands for a proposition that makes use of structural or functional concepts. However, *p* is mostly taken to be a proposition about an actual state of affairs that can be perceived directly (or a past state of affairs that can be remembered). A paradigm example of *p* is 'This object is red'. For this type of proposition, condition 1 is interpreted with the help of some theory of truth (usually the correspondence theory), and with regard to condition 3, visual perception is taken to be the source for justification. When *p*, however, refers to functional properties, then it becomes questionable whether or in what sense *p* still describes an actual state of affairs, since functional properties are often interpreted in terms of dispositional properties, which in turn are interpreted in terms of potential state of affairs (cf. Mumford 1998). The standard account of knowledge seems not geared to interpreting functional claims like 'This object is a broken steering wheel'.

From the point of view of engineering practice, but also from the point of view of daily life, it is hard to deny that we have *functional knowledge*. Engineers have knowledge of the functional properties of the components of a system and of the system as a whole; otherwise they would not be able to design, make and operate artefacts. They have knowledge of what things are for. Edison's description of the electric lamp, for example, contains a lot of functional knowledge about (parts of) the lamp. As regards daily life, Searle rightly remarks that as children 'we learn to perceive and use cars, bathtubs, houses, money, restaurants', that is, we learn to *perceive* [note Searle's use of language] things with functions and indeed, 'in most cases it is harder to see objects as just natural phenomena, stripped of their functional roles, than it is to see our surroundings in terms of their socially defined functions' (*The Construction of Social Reality*, p. 4) (Searle 1995). So the proposition *p* 'This object is a coffee cup' (i.e. 'This object has the functional property of being a coffee cup') appears to be a valid candidate for knowledge, just as the proposition 'This object is red'. Of course, to become knowledge, certain conditions with regard to *p* will have to be fulfilled.

If knowledge of functions is considered to be of the same kind as knowledge of structures, then the three conditions of the standard tripartite analysis have to apply. But with regard to condition 1, what does it mean for *p* to be true in case *p* is a functional proposition? Adherents to some versions of the correspondence theory of truth appear to get into trouble here because this theory implies that functions exist in a mind-independent reality. That, however, is a highly controversial point of view

---

[7] In spite of the criticism that has been raised against this account, it is still of paramount importance in epistemology; we will not enter into the discussion whether or not this criticism can be met by simply introducing extra or altering some of the conditions stated.

when it comes to technical functions; they are usually taken to be mind-dependent features of technical artefacts. And with regard to condition 3, is it indeed the case, as Searle says, that we (directly) perceive functions? If so, then perception can be the basis for justifying *p*, but if not, then we have to ask how the belief in *p* can be justified. [It seems difficult to maintain that functions can be perceived directly; archaeologists have a hard time reconstructing the functions of human artefacts.] At the background of these issues lie questions about what kind of properties functions are and what it means to say that an object has a particular function.

Let us move on to the third type of knowledge, that is, *prescriptive knowledge*. If we take technical action to be about changing our physical environment for some practical purpose, then it is rather obvious to assume that at least part of technical knowledge will have a prescriptive or procedural character. It will be about how to realize in an efficacious way changes in the physical world. That, of course, also presupposes descriptive knowledge about the actual state of the world and about the desired state. But an important element of technical knowledge is the description of procedures, that is, of prescribed actions that, when performed adequately, will bring about the intended change or state of affairs. This is immediately clear from Edison's patent; part of his invention is precisely a method of how to make carbon conductors of high resistance. This is knowledge he has developed and for which he hopes to secure propriety rights. With regard to the use of technical artefacts, this prescriptive nature of technical knowledge is quite familiar from the user manual: it tells the user what to do so that he may expect that the technical artefact will function properly. The same applies with regard to the making and repairing of technical artefacts. To know how to make or repair a technical artefact means knowing what series of actions has to be performed.

Prescriptive knowledge not only plays a dominant role in technological practice but also is of crucial importance in many other domains of human life, such as the social domain (law). Prescriptive knowledge does not fit in to the standard account because condition 1, about the truth of *p*, does not make sense when proposition *p* is a prescription or a set of prescriptions. Prescriptions may be effective (adequate) but not true. Within technology, these prescriptions are always hypothetical imperatives (conditioned on the desired change or state of affairs). It is also questionable whether prescriptive knowledge satisfies condition 2. A rule may be accepted or rejected in action, but it is not clear what it means to believe a rule. One may know (believe) that the application of a rule may be effective or not on the basis of previous (un)successful application of the rule and on the basis of this knowledge the acceptance or rejection of a rule may be justified. So for prescriptive knowledge, condition 3 may be formulated in terms of justification of the acceptance of a rule. All in all, this leads to the following tripartite analysis of A's prescriptive knowledge how to *p*:

1. *p* (i.e. *p* is effective).
2. A accepts *p*.
3. A's acceptance of *p* is justified.

According to this analysis, prescriptive knowledge is not a form of belief of a proposition expressing a state of affairs but of acceptance of a rule.

The foregoing shows that prescriptive knowledge is a form of know-how. But in order to avoid confusion, two *different* forms of know-how have to be distinguished:

- Know-how in conceptualized form, that is, know-how involving the explicit acceptance of a rule (or a set of rules) that describes how to proceed in order to reach a certain goal. This is this kind of know-how analysed above; it involves a proposition stating the rule (or the set of rules) (e.g. a user manual). Note that this kind of know-how (knowledge how to *p*) presupposes knowledge of the rule *p*!
- Know-how as skills (Ryle); this is a form of non-propositional knowledge in the sense that it does not involve a proposition stating a rule; it is even questionable whether this form of knowledge is rule-based and whether it can be rendered completely in a conceptualized form.

In the philosophical literature, know-how is usually associated with the second type and is opposed to knowing that, that is, propositional knowledge. The first type of know-how is also not a form of propositional knowledge, since it is a form of knowing how to *p* and not knowing that *p*. However, as we remarked above, it involves a form of knowing that, namely, knowledge that *p* where *p* stands for the proposition that '*p* is the rule'. This propositional element of the first kind of know-how is a form of knowing that which may be communicated through language. It plays an important role in technological knowledge. Just to mention one example, the editorial focus of the journal *Chemical Engineering Progress*, which is the flagship publication of the American Institute of Chemical Engineers, is to 'publish impartial, generic articles that provide some type of "how-to" guidance to the practising chemical engineer'.

Combining the distinction between structural and functional concepts and descriptive and prescriptive propositions, the following kinds of propositions may be distinguished:

	Descriptive	*Prescriptive*
Structural	This substance boils at 100°C	Add NaCl in order to raise boiling point
Functional	This is a thermometer	Open the steam valve in order to lower pressure

As far as technological knowledge is concerned, all four kinds of propositions may be involved in technological knowledge. Scientific knowledge is primarily associated with descriptive-structural propositions although in biology descriptive-functional propositions also play a role. Epistemologists have focused their attention on the analysis of descriptive-structural propositions, which has led to the standard conception of knowledge as justified true belief. As we have tried to show in the foregoing, it is questionable and a point of further research for epistemologists whether this standard conception constitutes a fruitful point of departure for analysing forms of knowledge with regard to the other kinds of propositions.

Design knowledge is of special interest for the epistemology of technological knowledge because designing is usually taken to be the feature which distinguishes technology from science. Most technological artefacts are based on a design, where a design may be taken to be a kind of blueprint indicating how to build the artefact. But usually there is more to a design in the sense that it also contains implicitly or explicitly an explanation about why the proposed physical structure fulfils the intended function. In other words, the design contains a technological explanation. Technological explanations contain structural and functional propositions in some kind of argumentative structure. This 'know-why' is an integral part of the knowledge base of engineers; knowing why a particular structure fulfils an intended function or, in case of malfunctioning, why it malfunctions is of great importance for engineering practice. From an epistemological point of view, this know-why is a very interesting form of knowledge, because it connects two different types of descriptions of the world. In so far functional knowledge is problematic from a justified true belief point of view; design knowledge is too since it makes use of this type of knowledge.

The analysis of *know-how*, the last topic on our list, is not unfamiliar in epistemology. This is mainly due to Ryle's analysis of the distinction between knowing that and knowing how. As we remarked above, know-how in the sense of conceptualized prescriptive knowledge plays a pivotal role in technological knowledge. But the same is true for know-how in the sense of skills. Here, we move out of the domain of knowledge that involves propositions. Know-how in the sense of skills is neither to be found in technology textbooks nor in technological journals; it is to be learned in exercises in the laboratory and in practice. The nature of know-how is still under debate. Whereas Ryle maintained that it is a form of knowledge *sui generis* relative to knowing that, others have maintained that this form of knowledge can ultimately be reduced to a form of knowing that (see Stanley and Williamson 2001). If that is indeed the case, know-how may be a form of knowledge that fits in the standard justified true belief approach to knowledge.

Know-how as a form of technological knowledge is usually associated with *physical* skills, that is, the capacity to perform certain physical operations (comparable to physical skills such as swimming and riding a bike). But technological know-how also encompasses *mental (intellectual)* skills. Mental skills, for instance, are required for calculating on the basis of mechanical principles the stresses in a beam, in order to ascertain whether the beam will support a certain load. Generally speaking, there appears to be no significant difference in the role of mental skills in technology and science. Knowledge of the laws of Newton including a firm grip of their meanings is just not sufficient to master Newtonian mechanics. Knowing Newtonian mechanics also means knowing how to apply its laws in specific situations. This know-how is learned by practice and is a mental skill. Although know-how as practical skill is often put forwards as a form of knowledge specific for technology in comparison to science, this is wrong. A theoretical physicist may do well without physical skills, but that is not true for an experimental physicist. Although know-how as physical skill may play a more dominant role in technology than in science, there appear to be, from a principal point of view, no real differences between the roles of know-how in both domains: in experimental science as well as technology, know-how in the form of mental and physical skills is indispensable.

6 Bearers of Knowledge: Beliefs, Actions and Other Categories

So far, we have paid little attention to the second condition in the standard analysis of knowledge as justified true belief. This second condition says that for A to know *p,* it is necessary that A *believes p*. The final question that we would like to raise is whether beliefs are necessary for all forms of knowledge and if not, what other kinds of knowledge bearers may be distinguished. According to Dancy, the justified true account of knowledge only defines propositional knowledge (p. 23), that is, 'knowledge that *p*; it does not define knowledge by acquaintance as in "*a* knows James" nor knowledge-how, e.g. knowledge how to ride a bicycle, unless these can be shown to reduce to knowledge that'. Apparently these nonbelief-based types of knowledge are not considered to be very important forms of knowledge, since Dancy does not analyse these in any further detail. Also Audi presupposes a strong link between knowledge and belief as may be concluded from the following remark (p. 3): 'Knowledge would not be possible without belief justification….'

There is an important difference in stance towards the world between science and technology. The principle attitude of the scientist towards the world is that of a spectator, of a (passive) observer. This is even true when a scientist takes recourse to experiments; then he actively intervenes in the world, but only to create the right conditions to observe the world. This role of being a spectator is still to be found in the original meaning of the notion of a theory, generally regarded as the highest form of knowledge produced by science ['theorein' in ancient Greece meaning 'being spectator at a play']. Thus, theoretical knowledge is the knowledge of a spectator, and the strong focus of epistemology on this kind of knowledge has led Dewey to his complaint that it leads only to 'spectator theories of knowledge' (Dewey 1929). The standard tripartite analysis belongs to this type of theories of knowledge. The attitude of the craftsman/engineer is much more that of an actor; he/she intervenes in the world with the intention to change that world, for example, to add new kinds of entities (artefacts like bikes, houses, etc.) to it. To achieve his aims, he needs reliable knowledge about how to act such that the desired change is realized. It seems impossible to arrive at an adequate analysis of the nature of this kind of knowledge starting from a spectator point of view.

The question whether beliefs are the sole bearer of knowledge is especially relevant for the analysis of know-how. Let us focus on the relation between know-how in the sense of physical skills and actions and begin with clearing up some possible misunderstandings. Actions, of course, can be the *object* of propositional knowledge: the proposition *p* may describe a particular action by somebody, for instance, that John turned a switch at a certain moment in time. The fact that actions may be the object of knowledge does not raise any specific problems within the standard account. The same applies to knowledge of know-how: the proposition *p* may say that John knows how to ride a bike.[8] But analogously to the rule-based form of

[8] It may be questioned whether there is something special about the justification of this kind of propositional knowledge from the first person perspective: suppose John claims that he knows that he knows how to ride a bike; what is the role of perception and action in the justification of this claim?

know-how, the possibility of propositional knowledge *about* know-how as skill does not imply that know-how itself is of a propositional nature.

Know-how as a form of non-propositional knowledge [assuming that that is the case] does not imply that actions are constitutive of this form of knowledge, in the same way as beliefs are for propositional knowledge. The reason is that know-how is not itself an action but implies a capacity or an ability to act. If *A* knows how to *B*, where *B* stands for an action or a series of actions, then some kind of dispositional property appears to be attributed to *A*: under appropriate conditions *A* will intelligently perform *successfully* action *B* (or a series of actions *B*). This disposition (capacity to act) is taken to be a form of non-propositional knowledge.[9] With regard to this type of knowledge, efficacy and efficiency are prime epistemological values instead of truth (at first sight, the notion of truth does not even make sense with regard to know-how): successful action (in the sense of intelligent performance) is the criterion by which to judge whether (or to what extent) somebody indeed possesses a certain kind of know-how or not.

More generally, when the knowing subject changes role from being a spectator to being an actor in the world, from scientist to engineer, the question emerges whether other sources for the justification of knowledge claims than perception, memory, consciousness and reason become available. The most obvious suggestion seems to be to consider action itself as a source of justification. Can action be a *direct* basic source of justification, without mediation of perception, for instance, in case of non-propositional kinds of knowledge such as know-how? As we remarked above, the best justification for know-how appears to be an actual demonstration of this know-how, that is, the successful intelligent performance of a series of actions. Depending on the kind of know-how involved, the successful intelligent performance is related to the series of actions itself (e.g. know-how to walk) or to the state of affairs that is the result of this series of actions (know-how to sharpen a knife). At this point, however, we have to be careful since the perception of these actions, not the actions themselves, or the perception of the resulting state of affairs appears to justify know-how.

But is perception and not action really the source for justification of know-how? In the discussion of this question, it is vital to distinguish between two different levels of analysis, which involve two different types of knowledge:

1. Know-how itself
2. Knowledge about know-how

The first kind of knowledge is not propositional, the second is. The justification for both types of knowledge may be different, but further research need to show in what sense they are different. The justification of know-how in terms of successful actions or the outcomes of these actions, as described just above, does not concern

[9] It is interesting to note that Audi (p. 75) interprets beliefs as dispositions; this means that if propositional knowledge is taken to be some kind of belief (true, justified etc.), then propositional knowledge, just as know-how, is of a dispositional nature.

the justification of know-how itself, but the justification of knowledge claims about know-how, that is, claims of the type: person *x* knows that person *y* knows how to *z*. Since these claims are of a propositional nature, the fact that perception plays an essential role in justifying this kind of claim should come as no surprise. It is in complete accord with the standard analysis of knowledge in terms of beliefs.

7 Conclusion

We conclude from our discussion that the theory of knowledge needs to extend its scope in order to be able to give an account of types of knowledge that do not fit the justified true belief paradigm. Technological knowledge is a key example where the standard approach fails. Functional knowledge, prescriptive knowledge, design knowledge and know-how cannot be accounted for this way. Among the many issues that need to be addressed are the taxonomy of types of knowledge in engineering, the nature of justification of these types of knowledge, the question what types of knowledge bearers should we allow in our analyses and the right characterization of the normativity involved in technological knowledge. Obviously such an analysis will involve meta-epistemological issues. An interesting discussion of some of these issues with regard to technological knowledge may be found in Houkes (2009) and Meijers (2009).

Appendix: Edison's Patent

Source: US Patent Office. Taken from http://edison.rutgers.edu/

T. A. EDISON.
Electric-Lamp.

No. 223,898. Patented Jan. 27, 1880.

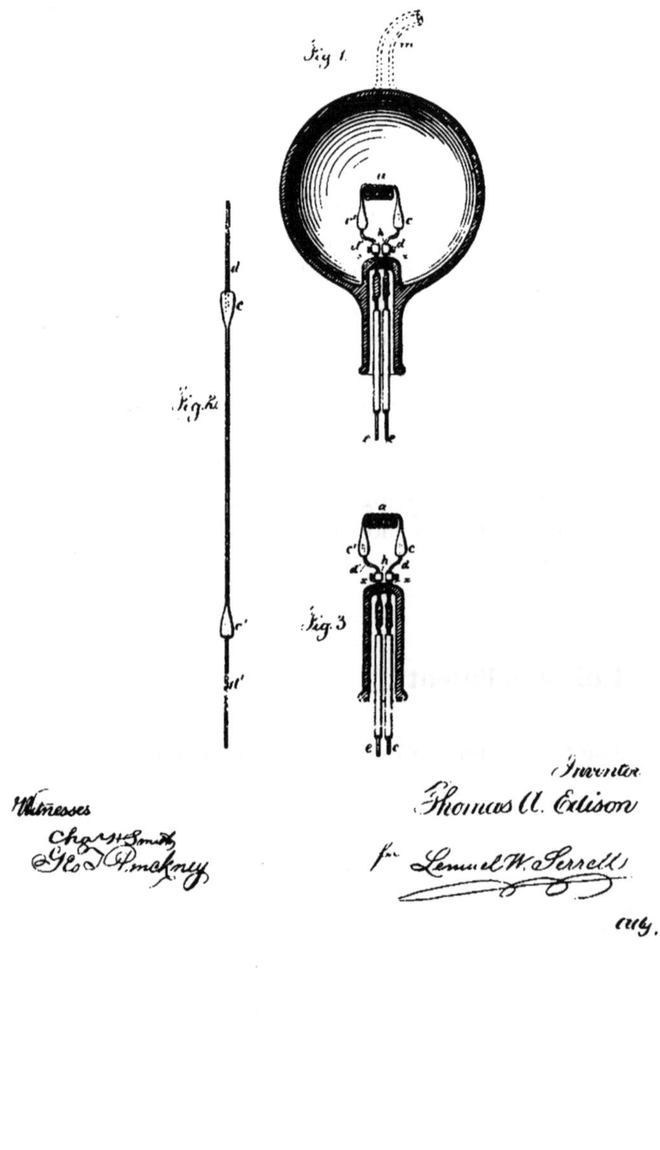

UNITED STATES PATENT OFFICE.

THOMAS A. EDISON, OF MENLO PARK, NEW JERSEY.

ELECTRIC LAMP.

SPECIFICATION forming part of Letters Patent No. 223,898, dated January 27, 1880.

Application filed November 4, 1879.

To all whom it may concern:

Be it known that I, THOMAS ALVA EDISON, of Menlo Park, in the State of New Jersey, United States of America, have invented an Improvement in Electric Lamps, and in the method of manufacturing the same, (Case No. 186,) of which the following is a specification.

The object of this invention is to produce electric lamps giving light by incandescence, which lamps shall have high resistance, so as to allow of the practical subdivision of the electric light.

The invention consists in a light-giving body of carbon wire or sheets coiled or arranged in such a manner as to offer great resistance to the passage of the electric current, and at the same time present but a slight surface from which radiation can take place.

The invention further consists in placing such burner of great resistance in a nearly-perfect vacuum, to prevent oxidation and injury to the conductor by the atmosphere. The current is conducted into the vacuum-bulb through platina wires sealed into the glass.

The invention further consists in the method of manufacturing carbon conductors of high resistance, so as to be suitable for giving light by incandescence, and in the manner of securing perfect contact between the metallic conductors or leading-wires and the carbon conductor.

Heretofore light by incandescence has been obtained from rods of carbon of one to four ohms resistance, placed in closed vessels, in which the atmospheric air has been replaced by gases that do not combine chemically with the carbon. The vessel holding the burner has been composed of glass cemented to a metallic base. The connection between the leading wires and the carbon has been obtained by clamping the carbon to the metal. The leading-wires have always been large, so that their resistance shall be many times less than the burner, and, in general, the attempts of previous persons have been to reduce the resistance of the carbon rod. The disadvantages of following this practice are, that a lamp having but one to four ohms resistance cannot be worked in great numbers in multiple arc without the employment of main conductors of enormous dimensions; that, owing to the low resistance of the lamp, the leading-wires must be of large dimensions and good conductors, and a glass globe cannot be kept tight at the place where the wires pass in and are cemented; hence the carbon is consumed, because there must be almost a perfect vacuum to render the carbon stable, especially when such carbon is small in mass and high in electrical resistance.

The use of a gas in the receiver at the atmospheric pressure, although not attacking the carbon, serves to destroy it in time by "air-washing," or the attrition produced by the rapid passage of the air over the slightly-coherent highly-heated surface of the carbon. I have reversed this practice. I have discovered that even a cotton thread properly carbonized and placed in a sealed glass bulb exhausted to one-millionth of an atmosphere offers from one hundred to five hundred ohms resistance to the passage of the current, and that it is absolutely stable at very high temperatures; that if the thread be coiled as a spiral and carbonized, or if any fibrous vegetable substance which will leave a carbon residue after heating in a closed chamber be so coiled, as much as two thousand ohms resistance may be obtained without presenting a radiating-surface greater than three-sixteenths of an inch; that if such fibrous material be rubbed with a plastic composed of lamp-black and tar, its resistance may be made high or low, according to the amount of lamp-black placed upon it; that carbon filaments may be made by a combination of tar and lamp-black, the latter being previously ignited in a closed crucible for several hours and afterward moistened and kneaded until it assumes the consistency of thick putty. Small pieces of this material may be rolled out in the form of wire as small as seven one-thousandths of a inch in diameter and over a foot in length, and the same may be coated with a non-conducting non-carbonizing substance and wound on a bobbin, or as a spiral, and the tar carbonized in a closed chamber by subjecting it to high heat, the spiral after carbonization retaining its form.

All these forms are fragile and cannot be clamped to the leading wires with sufficient force to insure good contact and prevent heating. I have discovered that if platinum wires are used and the plastic lamp-black and tar material be molded around it in the act of carbonization there is an intimate union by com-

2 223,898

material be molded around it in the act of carbonization there is an intimate union by combination and by pressure between the carbon and platina, and nearly perfect contact is obtained without the necessity of clamps; hence the burner and the leading-wires are connected to the carbon ready to be placed in the vacuum-bulb.

When fibrous material is used the plastic lamp-black and tar are used to secure it to the platina before carbonizing.

By using the carbon wire of such high resistance I am enabled to use fine platinum wires for leading-wires, as they will have a small resistance compared to the burner, and hence will not heat and crack the sealed vacuum-bulb. Platina can only be used, as its expansion is nearly the same as that of glass.

By using a considerable length of carbon wire and coiling it the exterior, which is only a small portion of its entire surface, will form the principal radiating surface; hence I am able to raise the specific heat of the whole of the carbon, and thus prevent the rapid reception and disappearance of the light, which on a plain wire is prejudicial, as it shows the least unsteadiness of the current by the flickering of the light; but if the current is steady the defect does not show.

I have carbonized and used cotton and linen thread, wood splints, papers coiled in various ways, also lamp-black, plumbago, and carbon in various forms, mixed with tar and kneaded so that the same may be rolled out into wires of various lengths and diameters. Each wire, however, is to be uniform in size throughout.

If the carbon thread is liable to be distorted during carbonization it is to be coiled between a helix of copper wire. The ends of the carbon or filament are secured to the platina leading-wires by plastic carbonizable material, and the whole placed in the carbonizing-chamber. The copper, which has served to prevent distortion of the carbon thread, is afterward eaten away by nitric acid, and the spiral soaked in water, and then dried and placed on the glass holder, and a glass bulb blown over the whole, with a leading-tube for exhaustion by a mercury-pump. This tube, when a high vacuum has been reached, is hermetically sealed.

With substances which are not greatly distorted in carbonizing, they may be coated with a non-conducting non-carbonizable substance, which allows one coil or turn of the carbon to rest upon and be supported by the other.

In the drawings, Figure 1 shows the lamp sectionally. *a* is the carbon spiral or thread. *c c'* are the thickened ends of the spiral, formed of the plastic compound of lamp-black and tar. *d d'* are the platina wires. *h h* are the clamps, which serve to connect the platina wires, cemented in the carbon, with the leading-wires *x x*, sealed in the glass vacuum-bulb. *e e* are copper wires, connected just outside the bulb to the wires *x x*. *m* is the tube (shown by dotted lines) leading to the vacuum-pump, which, after exhaustion, is hermetically sealed and the surplus removed.

Fig. 2 represents the plastic material before being wound into a spiral.

Fig. 3 shows the spiral after carbonization, ready to have a bulb blown over it.

I claim as my invention—

1. An electric lamp for giving light by incandescence, consisting of a filament of carbon of high resistance, made as described, and secured to metallic wires, as set forth.

2. The combination of carbon filaments with a receiver made entirely of glass and conductors passing through the glass, and from which receiver the air is exhausted, for the purposes set forth.

3. A carbon filament or strip coiled and connected to electric conductors so that only a portion of the surface of such carbon conductors shall be exposed for radiating light, as set forth.

4. The method herein described of securing the platina contact-wires to the carbon filament and carbonizing of the whole in a closed chamber, substantially as set forth.

Signed by me this 1st day of November, A. D. 1879.

THOMAS A. EDISON.

Witnesses:

S. L. GRIFFIN,

JOHN F. RANDOLPH.

Order of Cancellation of Certificate of Correction of Letters Patent No. 223,898.

It is found that the following certificate has been attached to Letters Patent granted to Thomas A. Edison for improvement in "Electric Lamps," No. 223,898, dated January 27, 1880:

DEPARTMENT OF THE INTERIOR,

UNITED STATES PATENT OFFICE,

WASHINGTON, D. C., *December 18, 1883.*

In compliance with the request of the party in interest Letters Patent No. 223,898, granted January 27, 1880, to Thomas A. Edison, of Menlo Park, New Jersey, for an improvement in "Electric Lamps," is hereby limited so as to expire at the same time with the patent of the following-named, having the shortest time to run, viz.: British patent, dated November 10, 1879, No. 4,576; Canadian patent, dated November 17, 1879, No. 10,654; Belgian patent, dated November 29, 1879, No. 49,884; Italian patent, dated December 6, 1879, and French patent, dated January 20, 1880, No. 133,756.

It is hereby certified that the proper entries and corrections have been made in the files and records of the Patent Office.

This amendment is made that the United States patent may conform to the provisions of section 4887 of the Revised Statutes.

[SEAL.]

BENJ. BUTTERWORTH,
Commissioner of Patents.

Approved:
M. L. JOSLYN,
Acting Secretary of the Interior.

Now, in compliance with the request of the parties in interest, said certificate is hereby *canceled* and proper entries and corrections have been made in the files and records of the Patent Office.

In testimony whereof I have hereunto set my hand and caused the seal of the Patent Office to be affixed, this 15th day of March, 1893.

W. E. SIMONDS,
Commissioner of Patents.

Approved:
CYRUS BUSSEY,
Assistant Secretary of the Interior.

Correction in Letters Patent No. 223,898.

DEPARTMENT OF THE INTERIOR,

UNITED STATES PATENT OFFICE,

WASHINGTON, D. C., *December 18, 1883.*

In compliance with the request of the party in interest, Letters Patent No. 223,898, granted January 27, 1880, to Thomas A. Edison, of Menlo Park, New Jersey, for an improvement in "Electric-Lamps," is hereby limited so as to expire at the same time with the patent of the following named, having the shortest time to run, viz: British Patent dated November 10, 1879, No. 4,576; Canadian Patent dated November 17, 1879, No. 10,654; Belgian Patent dated November 29, 1879, No. 49,884; Italian Patent dated December 6, 1879; and French Patent dated January 20, 1880, No. 133,756;

It is hereby certified that the proper entries and corrections have been made in the files and records of the Patent Office.

This amendment is made that the United States Patent may conform to the provisions of Section 4887 of the Revised Statutes.

BENJ. BUTTERWORTH,
Commissioner of Patents.

Approved:
M. L. JOSLYN,
Acting Secretary of the Interior.

References

Audi, R. 1998. *Epistemology; a contemporary introduction to the theory of knowledge*. New York: Routledge.

Dancy, J. 1985. *Introduction to contemporary epistemology*. Oxford: Basil Blackwell Ltd.

Dewey, J. 1929. The quest for certainty; a study of the relation of knowledge and action. In John Dewey the Later Works, 1925–1953: 1929/the quest for certainty by Jo Ann Boydston (Editor), Stephen Toulmin (introduction) and John Dewey.

Houkes, W.N. 2009. The nature of technological knowledge. In A.W.M. Meijers (ed.) 2009.

Lemos, N. 2007. *An introduction to the theory of knowledge*. Cambridge: Cambridge University Press. 232 pp.

Martin, A.L. (ed.). 1998. *Epistemology: The big questions*. Oxford: Basil Blackwell.

McAllister, J.W. 1997. Laws of nature, natural history, and the description of the world. *International Studies in the Philosophy of Science* 11: 245–258.

Meijers, A.W.M. (ed.). 2009. *Philosophy of technology and engineering sciences*, 1st edn. (*Handbook philosophy of science*), vol. 9, 1453 pp. Amsterdam: North Holland

Mitcham, C. 1994. *Thinking through technology; The path between engineering and philosophy*. Chicago: University of Chicago Press (Chapter 8: "Types of technology as knowledge").

Morton, A. 1997. *A guide through the theory of knowledge*. Oxford: Basil Blackwell.

Mumford, S. 1998. *Dispositions*. Oxford: Oxford University Press.

Polanyi, M. 1978. *Personal knowledge*. London: Routledge (1958, 1st edn.).

Pollock, J.L. 1986. *Contemporary theories of knowledge*. Totowa: Rowman and Littlefield.

Pryor, J. 2001. Highlights of recent epistemology. *British Journal for the Philosophy of Science* 52: 95–124.

Searle, J.R. 1995. *The construction of social reality*. New York: The Free Press.

Sosa, E., and J. Kim (eds.). 2000. *Epistemology; an anthology*. Oxford: Basil Blackwell.

Stanley, J., and T. Williamson. 2001. Knowing how. *The Journal of Philosophy* 97: 411–444.

Vincenti, W.G. 1990. *What engineers know and how they know it; Analytical studies from aeronautical history*. Baltimore: John Hopkins University Press (Chapter 7: "The Anatomy of Engineering Design Knowledge").

Chapter 3
Rules, Plans and the Normativity of Technological Knowledge

Wybo N. Houkes

1 Introduction

It is not uncommon to describe the efforts of engineers, the results of these efforts and the specific knowledge employed[1] in normative terms. Take, for instance, 'The engineer, and more generally the designer, is concerned with how things ought to be – how they ought to be in order to attain goals, and to function' (Simon 1981: 7).

Many such 'norm-oriented' characterisations are intended to be contrastive. In particular, they are supposed to differentiate designing, technology and technological knowledge from their counterparts in science by characterising the latter in non-normative, descriptive terms. For example, '(…) The modal mood of a pure scientist is largely descriptive, while the mood of engineering is generally prescriptive' (Hendricks et al. 2000: 278). Here, science is allegedly descriptive because it is aimed at truth or empirical adequacy. By contrast, the 'mood' or 'concern' of engineering is said to be at least partly prescriptive or normative,[2] because engineering is aimed at changing reality.

Critical examinations of norm-oriented characterisations of technology and engineering have mainly focussed on the proposed contrast with science – in line with the

[1] In this chapter, the efforts of engineers are labelled *designing*, the results of their efforts are *technologies* or *artefacts*, and the specific knowledge employed is *technological knowledge*. Occasionally, efforts, results and knowledge are indiscriminately referred to as 'technology', employing the ambiguity of this notion, which may refer to a system of knowledge, a system of objects (Mitcham 1978) and a set of (institutionalised) practices – as in usage of the term 'information technology'.

[2] Here and throughout this chapter, I take 'prescriptive' as a synonym of 'normative', that is, as 'concerned with norms or directives'. Thus, prescriptive statements include (expressions of) rules, recommendations and requirements.

W.N. Houkes (✉)
Eindhoven University of Technology, Eindhoven, the Netherlands
e-mail: w.n.houkes@tue.nl

M.J. de Vries et al. (eds.), *Norms in Technology*, Philosophy of Engineering and Technology 9, DOI 10.1007/978-94-007-5243-6_3,

once prominent debate on the science-technology distinction (Staudenmaier 1985). Elsewhere, it has been argued that this contrastive way of characterising technology may be counterproductive for understanding technology on its own terms, since any contrastive definition invites arguments that address the contrast rather than the characterisation of technology (Houkes 2009). One set of counterarguments might draw on similarities (in content, method and/or social institutionalisation) between science and technology. Another might start from the existence of intermediate activities (such as experimental science, applied science and engineering science). A third may be fuelled by (instrumentalist or empiricist) objections against characterising science as a truth-seeking activity. All these counterarguments undermine the supposed contrast between science and technology and make a characterisation of the latter guilty by association.

Although technology is frequently characterised in normative terms, papers in which the normativity involved is specified or analysed are few and far between. Yet it is easy to show that care is required in this specification, in particular if one is ultimately interested in contrasting science and technology. Suppose one takes a 'prescriptive mood' as essential for technology, in the sense that presence of this mood is indicative for technological rather than scientific knowledge. Then, one would be forced to conclude that science is free of prescriptive statements (the 'scientific mood' is descriptive), which is evidently false. In science handbooks, one can find prescriptive statements like 'It is *useful* to keep track explicitly of the total fields propagating in the two directions', 'Because of the generality of the contribution from the shadow region, it is *desirable* to consider it separately' (Jackson 1975: 372, 448; emphases added). Alternatively, one might maintain that, although one finds prescriptive statements both in science and technology, their role is different. This, however, requires a further specification; it should be spelled out what the difference is and where it comes from. Typically, the literature has little more to offer here than repeat the mantra that science is about truth and technology about usefulness. That, however, flies in the face of even mildly antirealist views of science and of the obvious relation between true theories and successful interventions. Another possibility is to seek differences in the exact form of prescriptive statements in science and technology. In any case, further analysis is called for – and the earlier point applies that seeking to contrast science and technology through normative terms puts an additional burden on the analysis, hampering an attempt to understand technology on its own terms and our ability to assess existing analyses.

The aim of this chapter is to examine critically the scope, source and type of normativity in technological knowledge. It inquires whether prescriptive content can be identified in technological knowledge and, if so, what generates this content and how it is related to other bodies of (technological and non-technological) knowledge.[3]

[3] The aim must be phrased in terms of prescriptive *content* of knowledge to avoid triviality: on non-naturalist conceptions, *all* knowledge involves norms on beliefs and is therefore 'prescriptive'. The guiding idea behind this chapter is that, in addition to the norms on knowledge, there are norms *in* some knowledge, that is, it is about prescriptions.

The analysis is not biased towards capturing the 'essence' of technological knowledge or towards distinguishing it clearly and univocally from science or other domains of practice and inquiry.[4] Technological knowledge is here taken as a broad epistemic category, encompassing the knowledge engineers employ in practice, the knowledge users have regarding technologies and the supposedly more systematic and detailed bodies of knowledge in the engineering sciences. Thus, technological knowledge is likely to be internally diverse, requiring a taxonomical rather than monolithic approach. This chapter examines whether, among the subtypes of technological knowledge, there is a sufficiently robust prescriptive subtype.

The relevance of this examination lies not in the need to revive the science-technology debate, nor does this chapter stem from the conviction that epistemological issues regarding technology or the engineering sciences are a burgeoning topic of philosophical inquiry. There are at least two areas outside of philosophy, however, where appeals to technological knowledge are frequently made and in need of further analysis. The first area is that of methodological inquiries into what might be called 'intermediate' disciplines, which combine scientific theorising with striving for specific practical relevance. Examples of these disciplines are management and organisation studies. It has been proposed to understand these disciplines as design sciences, aimed at producing a particular kind of prescriptive, technological knowledge – namely, technological rules. This proposal explicitly builds upon literature that has, to put it mildly, become suspect in the philosophy of technology – most notably, on Mario Bunge's work on technology as applied science. The other area in which technological knowledge plays a prominent role is evolutionary economics. There, several models of organisational behaviour and development (Nelson and Winter 1982; Cohen et al. 1996) have been presented in which knowledge or 'routines' within organisations are the prime unit of analysis or even the genetic basis of firm evolution.

The inquiry starts by presenting Mario Bunge's (1966) concept of a technological rule and identifying its shortcomings with respect to the aim of this chapter (Sect. 2). These shortcomings are then amended by means of the plan analysis of using, designing, making and manufacturing, proposed by Pieter Vermaas and the author (Sect. 3). After finding, in this analysis, sufficiently robust prescriptive content in technological knowledge (Sect. 4), the role of knowledge in evolutionary economics is chosen as an area of application. It is argued that analysing technological knowledge within organisations in terms of knowledge of plans or procedural knowledge, rather than exclusively in terms of routines and tacit knowledge, may enrich and substantiate models of technological change developed in evolutionary economics (Sect. 5).

[4] The aim is not even to capture the only or the only important prescriptive part of technological knowledge. Another part, not discussed in this chapter, consists of the category-specified value statements (e.g. 'This is a good bridge') identified by Hansson (2006).

2 Technological Rules and Norms

Above, the difficulties in determining the prescriptive content of technological knowledge have been briefly indicated. Rather than ex ante assuming a contrast with science, one needs to analyse technological knowledge in its own right. One way of doing this is by introducing a central normative concept that, intuitively, captures at least part of technological knowledge and by explicating this concept as it applies specifically to technology.

Mario Bunge's analysis of technology as applied science in the eponymous paper (1966) and some later work offers such a concept – namely, that of a technological rule.[5] Of the many ideas Bunge develops in this paper, only his identification of technology and applied science has been discussed or, at least, vehemently rejected. That Bunge is not interested in downgrading the work of engineers to the mere application of existing scientific knowledge and that he, in fact, stipulates the technology-as-applied-science 'thesis' in order to simplify the discussion of other epistemic aspects of technology, has been largely overlooked.

Bunge introduces technological rules in the context of a dual contrastive exercise: in addition to the familiar science-technology contrast (here, a contrast between pure and applied science), Bunge wants to distinguish applied science from the knowledge employed by craftspeople. The pure-applied distinction is primarily specified in terms of goals of the (epistemic) practices and secondarily in terms of the content of the knowledge produced. The goal of applied science is practical rather than theoretical, meaning that it aims at 'enhancing our welfare and power'. In order to realise this goal, theories in applied science are 'the basis of a system of rules *prescribing* the course of optimal practical action' (1966: 330; emphasis added). Rules in general are then defined as 'instruction[s] to perform a finite number of acts in a given order and with a given aim' (1966: 338). Within this broad class, 'rules of science and technology' are one of four subtypes – the others being rules of conduct, rules of thumb in arts and crafts and rules of syntax and semantics. Rules of science and technology are characterised as 'grounded rules of research and action', where a rule is grounded if and only if it is 'based on a set of law formulas capable of accounting for its effectiveness' (1966: 339). Bunge contrasts grounded rules explicitly with conventional rules, which he takes to comprise the bulk of the first three subtypes. Finally, he implicitly discusses two ways of grounding technological rules,[6] namely, through substantive and operative technological theories. Both provide 'understanding of the modus operandi' of a rule, the first through providing knowledge regarding the objects of action, for example, machines, and the second by conceptualising action itself. Bunge submits that the former apply general scientific theories to particular types of practically relevant devices or processes, such as 'flight' or 'petrol drilling'. Operative theories, by contrast, are

[5] Bunge's later work differs marginally from Bunge (1966).

[6] Interestingly, Bunge does not discuss the 'rules of science' in his paper.

scientific in their method and not in their content: examples given are 'the theory of airways management' or of 'inventory practices' and, more generally, decision theory.

This brief presentation suffices to show that Bunge's analysis of technology is more detailed and varied than typically assumed. It may also be illuminating with respect to understanding the normativity of technological knowledge. Closer scrutiny, however, brings to light three main shortcomings for our present purposes. The first is that it presents an unduly restrictive view of the relation between prescriptions and more theoretical knowledge; the second is that it does not differentiate technological rules by their content but only by their underpinning; and the third is that it leaves unanalysed to which items the prescriptions apply, with which force and by which standards. Let us look at each shortcoming in some detail.

The first resembles, but broadens and refines, the standard objection levelled against Bunge that theories about artefacts and processes cannot generally be understood as mere applications of scientific theories (e.g. Vincenti 1990; Kroes 1992). As it stands, this objection only targets Bunge's characterisation of substantive technological theories. A broader problem is that Bunge insists on a relation between technological rules and sets of law formulas that is evidently too strict and offers no clear alternative. The standard objection a fortiori makes clear that no law formulas could have been involved in understanding the effectiveness of rules regarding early steam engines or airplane stability, since there were no substantive theories to ground these rules in. Moreover, it is unclear whether some of the examples Bunge gives of operative theories – such as decision theory and queuing theory – may serve as grounds for rules, since they are formal rather than lawlike, and, as formal systems, it is debatable how, if at all, they contribute to our understanding of the modus operandi of action sequences. Furthermore, even supposing that understanding the modus operandi of socio-technical systems like airways and their management would be possible on the basis of lawlike statements, it is not clear why a similar understanding would not be available for many rules of conduct: exchanging money for goods or protecting other members of a group is not obviously different from managing air traffic. This possible lack of differentiation between 'systems of actions' would also become clear if one would de-emphasise law formulas as the basis for rules in favour of, for instance, measures of success: many or most rules of thumb and rules of conduct would count as grounded, unless one would add specific glosses on the relevant notion of success for technological rules.

This relates to a second problem: even in their narrow, law-grounded specification, technological rules cannot be distinguished from rules in general by their content but only by their underpinning. Thus, the prescriptive statements in engineering textbooks would be of the same form as those in self-help books, although they are different in being grounded. To appreciate the width of the resulting class of rules (disregarding, for the moment, the first problem), consider dietary restrictions. These fall under Bunge's notion of rules, since they are instructions to prepare or eat foods in certain ways in order to, for example, reduce blood pressure. A diet that minimises sodium consumption consists of a prescribed (in this case also proscribed) series of actions. The effectiveness of these actions is well understood, and even the reason for prescribing the diet is based on knowledge of the composition of food, the physiological effects of sodium, etc. The same goes for rules such as 'If you are caught in a thunderstorm, avoid trees and large bodies of water and roll up in a ball'. These are even more evidently part of practical wisdom rather than technology, but based on knowledge of the natural world.

One cannot object to Bunge's analysis on the ground that he equates diets and engineering science: he intends the analysis to apply to medical technologies, including therapies and diets, and all kinds of behavioural repertoires. Some of his examples suggest that Bunge intends technological rules and thus technology to have a wide scope, encompassing every systematic pursuit of practical goals. Rules for action may then concern any action whatsoever. Simultaneously, he mentions 'use', 'design' and 'manufacture' as the subjects of operative technological theories – which suggests a more narrow understanding of technology. Only the latter is useful for our present purposes, but Bunge's paper offers no hint which specific rules may apply to narrowly technological actions such as use and design.

These two problems concern identification of the prescriptive content of technological knowledge. The third and final concerns the characteristics (source, locus and force) of the prescriptions involved. Bunge rightly notes that technological rules vary from recommendations to requirements and strongly suggests that whatever force the rules have derives from the practical aim, that is, the aim of 'enhancing our welfare and power'. Thus, technological rules seem to derive their force from standards on practical rationality: given an aim, taking certain actions rather than others is advisable or even necessary. This is intuitively plausible but problematic to develop in the context set up by Bunge. For practical rationality applies primarily to actions, including individual actions, and only derivatively to instructions. Moreover, is an instruction rational if it recommends actions that are rational or if it requires them? And how are the individual actions prescribed? Rules typically take a linear form ('First do A, then B, then C, …'), whereas for many technological situations, actions are only partially ordered: you may take some actions in any order you like, as long as you have taken all of them before proceeding to taking another action. Conjoining all prescribed actions into a single 'instruction' leaves it unclear with what force the individual actions are prescribed, whether the series of actions is evaluated in addition to the individual actions, and how the instruction prescribes both the actions and the series. These aspects of evaluating actions cannot be incorporated without analysing the internal structure of a technological rule, that is, unpacking the various actions and their evaluative relations. Bunge's proposal does this only implicitly, if at all.

As said above, Bunge's work was recently picked up outside philosophy, in a discussion regarding the nature and method of so-called design sciences. Joan Van Aken (2004, 2005) has, in this context, introduced a modified form of Bunge's technological rules: 'a chunk of general knowledge linking an intervention or artefact with an expected outcome or performance in a certain field of application' (2005, 23). Like Bunge, Van Aken differentiates the contents of a design science – field-tested and grounded rules – in terms of its underpinning: field-tested means that a rule is tested in its intended field of application; grounded means that it is known why the intervention or artefact gives the desired performance.

This chapter is not concerned with the nature of design sciences, so no evaluation of the modified characterisation with respect to that goal is given here. Yet with respect to the goal of identifying prescriptive content in technological knowledge, Van Aken's technological rules share the calibration problems noted for Bunge's: only a small class of interesting technological knowledge is grounded in a strict sense, and the wider sense suggested by the term 'linking' generates a broad understanding of technology. The reason is that, again, groundedness is introduced as the only distinctive feature and that the notion cannot bear this analytical burden. Van Aken refers to artefacts in his general characterisation. Since this, however, is a disjunction, it expresses the idea that some technological rules involve only interventions by means of the human body or natural resources. Bunge's third problem, of the prescriptive character and the internal structure of rules, is exacerbated on Van Aken's modification: it may not be clear how standards on action translate into standards on instructions to take series of actions, but far less clear how such standards affect 'chunks' of general knowledge. Moreover, in referring to 'expected' outcomes and omitting references to practical goals, Van Aken decreases the prescriptive force of technological rules to that of (theoretically) rational expectations. Newton's law of gravitation links shoving a book off the table to its falling to the ground, which is the scientifically well-understood and expected outcome. Calling this law a technological rule deprives the latter notion of interesting prescriptive content.

3 Plans and Agents

The previous section outlined some literature that may be of use in identifying and analysing the prescriptive content of technological knowledge. Three questions were raised in this regard, which may be summed up as follows: prescriptions concern actions, but (a) how are prescriptions on actions, again in a technological context, related to other knowledge, in particular more theoretical knowledge? (b) to which types of actions do these prescriptions apply in a technological context? and (c) on the basis of which standards, with what force and at which level of aggregation are these actions prescribed?

It was argued that analyses in the literature provide implicit, incorrect, very partial or even no answers to these questions. Switching to a more constructive mode, this section will present the main elements of the plan analysis of a number of technological actions developed by Pieter Vermaas and myself. In the next section, it is argued that this analysis provides answers to the critical questions listed above.

The plan analysis was initially developed for reconstructing using and designing (Houkes et al. 2002; Houkes and Vermaas 2010, Ch.2) and later extended to making and manufacturing (Houkes and Vermaas 2009). It builds on an approach in the theory of action in which plans rather than individual intentions are considered as the products of practical deliberation (Bratman 1987; Pollock 1995). More importantly, plans are taken as the units of evaluation: in assessing practical rationality, one should consider courses of intentional action. We characterise plans as complex items that consist of considered actions; the actual actions constitute a plan execution rather than the plan itself. This execution is a physical process that involves the human body and possibly other material objects. We refer to the mental process of planning as constructing a plan. This process results in a more or less enduring mental state, similar to a belief or intention, and different from a fancy. Plans can also be reconstructed, for example, when assessing the rationality of actions. Finally, all plans crucially involve the following elements, which serve to differentiate plans from each other:

(a) A goal state to be realised
(b) An (partial) ordering of considered actions
(c) A set of objects used

where (c) is a quick-and-dirty way of identifying (b). Both (b) and (c) identify 'means' by which the 'end' (a) is achieved, since 'means' may refer to both goal-directed actions and the objects manipulated in these actions (Hughes 2009).[7]

Within the broad class of plans, we distinguish use plans and make plans. If some of the considered actions are manipulations of artefacts, we call the plan a use plan; if the goal of the plan is to produce an object, we call the plan a make plan. Some plans are both make plans and use plans, since their aim is to produce an object and they involve the manipulation of tools, that is, other items than the object to be produced.

To illustrate some features of use plans, consider the goal to lose weight. One may consider various courses of action for achieving this goal. Some of those only concern one's own body – for instance, walking to and from work instead of taking the car. Other plans may involve other objects, such as food: one may consider breakfasting on oatmeal instead of donuts for four weeks. Still other plans involve objects that are uncontroversially technical artefacts: one may consider undergoing liposuction. Some of these plans, such as (hopefully) the latter, also involve other people.[8] One may, but typically does not, construct a plan from scratch; most actual deliberation involves choosing between plans that are well known in a society (e.g. from diet books, commercials or hearsay). Alternatively, one may tailor-make a plan to one's own situation or preferences (e.g. by substituting cauliflower with broccoli in a suggested diet). Virtually all plans consist of more than one action.

---

[7] Here, the analysis connects to the notion of 'practical norm' introduced by Von Wright (1963) and, indirectly, to the analysis of instrumental knowledge given by Niiniluoto (1993). Like Bunge, Niiniluoto aims at understanding applied research through focussing on relations between means (actions) and ends. His discussion and its relation to the present analysis are complicated, however, by Niiniluoto's insistence on distinguishing between technology, which concerns actions and is evaluated in terms of its effectiveness, and instrumental knowledge, which concerns technical norms and is evaluated in terms of truth. This raises the question whether technical norms are indeed truth candidates (an issue Von Wright stays agnostic about) and the question how technology and instrumental knowledge are related, although they involve different core values. I submit that the plan analysis clarifies the relation while restoring Von Wright's agnosticism but will not substantiate this here.

[8] The plan analysis does not systematically distinguish between contributions to goal realisation by objects and by other people. Like plans may require artefacts to manifest dispositions (e.g. to conduct heat or electricity), they may require actions taken by people (e.g. drive your train).

Sometimes, the order of these actions matters (e.g. anesthetising before undergoing liposuction); sometimes, the order is arbitrary (e.g. a diet for one week may contain seven recipes for dinner courses without requiring a day-to-day order). Finally, the actions considered may be very elementary (e.g. walking) or may require great skill and 'environmental support' in their execution, to the point that they require additional (sub-)planning.

Several general advantages of focussing on plans rather than individual actions have been brought forward, such as the temporal stability of plans and the possibility of making sense of 'locally detrimental' actions. Our plan analysis[9] was constructed for different purposes – namely, to analyse the design and use of artefacts in terms of intentions and to identify those intentions that are relevant for ascribing functions to artefacts.

Using an artefact may be analysed as executing a use plan that includes at least one manipulation of the artefact. Designing is reconstructed as constructing and communicating a use plan and, subsidiary to that, as describing the types of items manipulated in the plan. In this analysis, designers primarily aim at aiding prospective users to realise their goals.[10] Central to such assistance is developing a sequence of actions to be undertaken by users and communicating it to them via, for example, user manuals, explicit instruction and features of the artefact, against a background of known habits and cultural patterns. The considered actions constituting a plan and their ordering can be communicated verbally: if an agent who knows how to realise a certain goal tells another how he went about realising it, he communicates a series of actions. Communicating this 'procedural' aspect of artefact use does not, of course, immediately give the other agent the capacity to realise the goal, since some or all steps in the plan may require skills that the other agent does not possess. This shows that actions cannot be described in terms of plans alone. Analysing and especially evaluating the role of skills and external circumstances is, however, possible on the basis of reconstructing goal-directed behaviour as plan execution (Vermaas and Houkes 2006). Whether exercising your skill to, for example, ride a bicycle is rational depends on the role of this skill in a larger plan, for example, to cycle to work. There may be other ways of evaluating skills and their exercise, but one evaluates them as means to a practical end.

On the plan analysis, 'user' and 'designer' are roles that may be played by different agents, in what might be called a 'division of practical labour'. Designing is aimed at a second-order goal, namely, of contributing to the realisation of practical goals. Constructing and communicating a use plan is the designer's way of realising this central goal. This is a broad notion of designing, including the activities of therapists

[9] The general merits and difficulties of a plan-centred approach are not considered here. Planning may or may not, for instance, be reducible to a series of intention formations; this is irrelevant for present purposes as long as the (possibly irreducible) concept of use plans can be used to analyse artefact design and use, and the prescriptive content of technological knowledge.

[10] This characterisation of designing resembles Herbert Simon's: 'Everyone designs who devises courses of action aimed at changing existing situations into preferred ones' (1981, p.129).

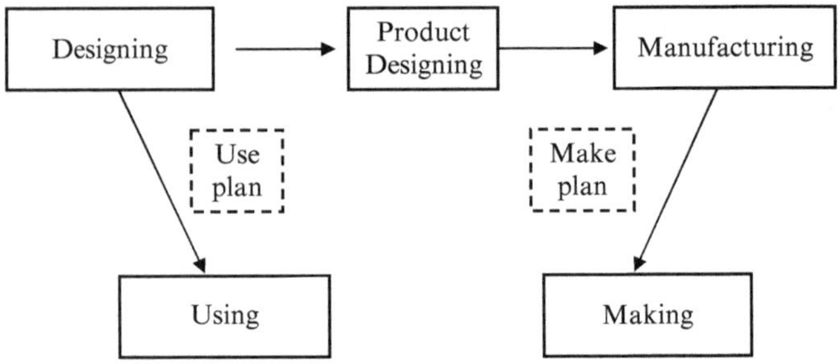

Fig. 3.1 Agent roles and communication in the plan analysis

and consultants. Only the subsidiary notion of product designing may distinguish the activities of (some) engineers from those of other 'practical aides'.

The plan analysis may be extended to other artefact-related activities, most notably production. There, making an artefact can be characterised as executing a make plan: a series of considered actions, possibly (and typically) including the manipulation of one or more tools, aimed at producing an object. The role of constructing and communicating this make plan can be conceptually distinguished from other roles. Following a division of labour that is characteristic of contemporary engineering, this fourth role may be labelled 'manufacturing' (Houkes and Vermaas 2009). The resulting analysis of agent roles looks like that in Fig. 3.1. Two pairs of activities, using-designing and making-manufacturing, are related by communication of plans, use plans and make plans, respectively. Designing and manufacturing are related by exchange of information, via product designing: designing results in a specification of objects to be manipulated while executing a use plan, including the capacities by which these objects contribute to realisation of the goal of this plan (the objects' functions); product designing develops this specification of an object into a blueprint, including the functional and structural decomposition of the object.

One advantage of the plan analysis is that it differentiates the agent roles involved in technology, identifies the goals and actions involved in playing these roles, and thereby relates the various roles. Another advantage is that it provides a framework for evaluation. The cornerstone of this framework is that all four roles can be assessed on the basis of the quality of the plan that is executed or constructed, relative to the circumstances in which it is executed or constructed. In ordinary language, plans may carry such normative labels as 'realistic', 'sound' or 'ruthless'. Here, effectiveness is taken as the core quality of a plan: a plan is only of value if agents are likely to realise the corresponding goal by executing the plan.[11] Effectiveness is typically judged in combination with efficiency – a comparative value, where the reference class is determined by available alternatives, circumstances and skills of

[11] The effectiveness standard on plans may be so fundamental that the very notion of 'plan' incorporates a measure of effectiveness. If I would try to find extraterrestrial life by reading *The Waste Land* in my roof garden, I would not be executing an irrational plan, but no plan at all (e.g. because my house has no roof garden).

an agent. One may deliberate whether to take the car or train to work, unless one does not have a driving licence; methods for curing diseases should improve upon existing methods in terms of cure rate or side effects; and household items are often sold as less time-consuming, costly or waste-producing. A plan is here called '(practically) rational' if it is effective and efficient.

Other standards for or demands on plans that have been proposed in the literature concern goal, means-end and belief consistency – which all concern part of the internal structure of plans.[12] If plans are directed at multiple goals, or have subsidiary aims, the agent executing the plan must reasonably believe that both goals can be realised by executing the plan; if not, the plan is goal inconsistent. A plan to phone while driving home is goal inconsistent if the driver knows that this leads to risks she/he finds unacceptable (taking safety as a subsidiary goal). Means-end consistency concerns the possibility to execute all considered actions, given the availability of auxiliary items, the agent's competences and environmental conditions. This standard partly explains the typical dynamics of constructing plans. The plan to drive a car cannot be executed successfully if, among other things, the tank is empty, the tyre pressure is too low or the windows are smeared with mud. Hence, a driver has to stop at a gas station, garage or carwash every now and then. Finally, plans are based on beliefs about the world, ourselves and the effects of the actions. We require not just that the use or make plan be based on actual beliefs but on justified or reasonable actual beliefs. Requiring true beliefs, or knowledge, would be too strict: even seemingly flawless deliberating practices are not guaranteed to lead to effective plans. Someone may, for instance, justifiably believe that he can cycle from Amsterdam's Central Station to the Van Gogh Museum via the Dam square, only to find that the Dam square is blocked because of (unannounced) construction work. Intuitively, there is little wrong with the process that led to this travelling plan, except that executing the resulting plan does not lead to realisation of the goal.

Since plans can be evaluated in terms of their rationality and use can be described as the execution of a use plan, using an artefact is rational if and only if it is the execution of a rational use plan. Standing on a chair to change a light bulb is rational, provided, among other things, that one justifiably believes that the chair can hold one's weight (i.e. that the executed use plan is effective) and that there are no artefacts available that are more appropriate to stand on (i.e. that the executed use plan is efficient). Thus, the beliefs and specific circumstances of individual users, for example, concerning their weight, determine whether artefact use, considered as a particular event, is rational or irrational. This distinction between rational and irrational use must be distinguished from that between proper and improper use, which refers to social institutionalisation of use and use plans (Houkes and Vermaas 2010, Ch.2). The evaluation of making builds in the same way on the quality of the make plan executed.

[12] These standards are taken from Bratman (1987, §3.2), who uses 'internal consistency' for the first standard. We changed this label to indicate that the other standards also derive from the internal structure of plans. See also Audi (1989, Ch.7).

Since the goal of designing and manufacturing is to contribute to other activities, the effectiveness of the activities is determined by the success of the contribution: if a designer constructs an irrational use plan and thereby does not aid users in realising their goals, or if he does not effectively communicate a rational plan to prospective users, he has not achieved his own goal and designing is irrational. This, by the way, leaves open the possibility that (product) designers ground their work in models that are known, also by the designers themselves, to be incorrect. As long as their belief in the effectiveness of the communicated use plan is justified, for example, on the basis of extensive tests, such designing can be evaluated as rational. Again, since manufacturing is the 'productive counterpart' of designing, it is evaluated in the same way, substituting make plans for use plans.

4 Normativity in Technological Knowledge

In this section, it is shown that and how the plan analysis provides a basis for identifying and analysing prescriptive content in technological knowledge. This result captures some elements of the proposals discussed in Sect. 2 while avoiding the three problems identified there.

In particular, the plan analysis employs the idea that technology involves 'useful' knowledge, that is, knowledge for producing artificial systems and using these and natural systems in order to realise a practical goal. Following Ryle's well-known distinction between knowing-how and knowing-that, many authors have assumed that this 'know-how' constitutes a separate, non-propositional type of knowledge. This neglects, however, that useful knowledge can be analysed as having two components: knowledge that a sequence of actions leads to the realisation of a goal, that is, knowledge of a plan, and the skills needed to take these actions. The former may be called 'procedural knowledge' and the latter 'operational knowledge'.[13] This underwrites Ryle's original point that procedural knowledge does not exhaust instrumental knowledge, like the activities analysed in Sect. 3 cannot be described exclusively in terms of plans. Like one may know how to drive without being able to, one might possess procedural knowledge of driving without possessing the skills or means to realise the goal in practice. Likewise, one might communicate the former without thereby communicating the latter, for example, in handing over sheet music describing Chopin's preludes.

The standards on plans lead to a straightforward evaluation of procedural knowledge: person *p* only knows how to realise a goal if *p* knows a rational plan for realising this goal. Relatedly, person *p* knows how to use artefact *a* only if *p* knows at least one

[13] Calling knowledge of plans 'procedural knowledge' matches usage of the latter term in AI (Georgeff and Lansky 1987). The term is also used in cognitive psychology (e.g. Anderson 1981), where it indicates knowledge of how to realise a goal, that is, what we have called 'instrumental knowledge'; cognitive psychologists contrast procedural knowledge with declarative knowledge, which we would call 'descriptive'.

rational use plan *up* that includes manipulating *a*; and *p* knows how to produce or make artefact *a* only if *p* knows at least one rational make plan *mp* with the aim of bringing *a* into existence. To give some examples, Diane only knows how to become a millionaire if she is aware of at least one series of actions, execution of which would effectively and efficiently lead her to possess at least €1,000,000; this plan may be a rational (albeit illegal) plan to rob a nearby bank, or it may be a plan to sell her million-euro house. William only knows how to use a shaver if he knows one way to realise a goal by manipulating the shaver. This goal is, presumably, shaving by plugging in the shaver, switching it on, etc., but other goals and series of actions are conceivable (recall the distinction between rational and proper use). Terry only knows how to make an omelette if he knows one series of actions (in this case, a recipe) that reliably brings an omelette into existence. In many cases, these types of procedural knowledge are related. Alessandro, for instance, only knows how to make a cappuccino if he knows a series of actions that produces steamed-milk foam. The make plan for this foam involves a series of actions, including manipulation of a pitcher: knowing how to make foam includes knowing how to use the pitcher.

Through the standard on plans, procedural knowledge – and therefore instrumental knowledge – is intrinsically related to practical rationality and to practical reasons: having this knowledge provides reasons for doing something rather than just believing something (Franssen 2006). Conversely, in the absence of such practical reasons, one cannot claim to have procedural knowledge. Given a desire for cappuccino and the availability of a pitcher (and other resources), knowing how the pitcher can be used to make foam amounts to a reason for use: using the substance to fulfil one's desire is defeasibly rational, since one knows how it can be used successfully as such.

Knowledge of how to produce or manipulate objects and systems in order to realise a practical goal is an important part of technological knowledge, although it does not exhaust it. In turn, procedural knowledge is one component of instrumental knowledge. Since we found out that procedural knowledge is prescriptive, in being subject to standards of practical rationality, it follows that technological knowledge has a prescriptive element: it comprises knowledge of plans for manipulating or producing artificial systems, and this knowledge is primarily subject to standards of practical rationality. Practical rationality does not fully replace theoretical rationality in the evaluation of procedural knowledge, however, because a use or make plan is practically rational only if it is based on justified, that is, theoretically rational, beliefs. Theoretical rationality is, however, subordinate to practical rationality in this context: that the belief base of plans is subject to traditional epistemic norms is one of the standards of practical reasoning imposed on procedural knowledge.[14]

It is not difficult to find examples of such prescriptive or procedural elements in paradigmatic examples of technological knowledge. Textbooks in the engineering

---

[14] This touches upon a deeper worry regarding the status of procedural knowledge as knowledge, namely, whether knowledge of a plan is a candidate for being true. The standards of practical rationality may distinguish good and bad examples of this type, but this does not resolve whether it is a properly *epistemic* type.

sciences typically include procedures, regulations or recommendations regarding the design, control or manufacturing of technical systems and processes; examples from two arbitrarily chosen textbooks from different fields include instructions for permanent mould casting processes, V-bending and water jet cutting (Groover 2007, p. 227, 448 and 621, respectively) or for superheating in vapour power plants and optimising regenerative gas turbines (Moran et al. 2003, p. 198–199 and 243–245). In engineering practice, standards and regulations abound. Many of these may not stem directly from practical rationality but impinge on it because they concern the way in which practical goals may be pursued within a social or legal context. In use practice, prescriptive knowledge is most clearly expressed in user manuals, which prescribe series of actions for realising a goal. Their form closely resembles Bunge's technological rules, although the goal or desired state of affairs is often left implicit. In all cases, verbal expressions (or even pictorial, as in flow diagrams) of the procedural knowledge represent plans as linear orderings (e.g. by presenting them as an order of numbered steps), whereas the standards of rationality may only require a partial ordering of some steps (i.e. for effectiveness and efficiency, it does not matter which action is taken first). In all cases, skills and environmental support are presupposed in communicating plans – this communication only shows that relevant procedural knowledge has been identified and verbalised, not that this component outstrips operational knowledge, let alone that it exhausts instrumental knowledge.

As these cases show, procedural knowledge need not have the prescriptive force of requirements. Artefacts may be used and produced in many different ways, not all of which may or can have been envisaged by designers or manufacturers. The plan account incorporates this by a liberal notion of design – and, by implication, of manufacturing. Everyone, engineers and consumers alike, can construct and communicate plans for using and producing artefacts. One need not have a degree in engineering to use brown paper to remove candle wax from clothing – use that is as effective and efficient as it is widespread – nor to make a swing out of an old tyre and a piece of rope. Both activities may be based on instrumental knowledge, the procedural element of which can be made explicit. This knowledge does not carry the force of requirements or rules, especially in the tyre-swing case; instead, it involves a recommendation or practical suggestion in some circumstances: there is no need to make a swing, or a need to use a tyre in case one has decided to make a swing, or to tie particular knots in case one has decided to make a swing with a tyre. This point is easily put in terms of means: the actions included and the artefacts or instruments manipulated in the plans can be necessary means for realising a goal, but this is the exception rather than the rule for most technologies. Most actions and artefacts are, in the context of a use or make plan, sufficient means – where the availability of alternative means is common knowledge.[15]

The answers given in the plan analysis to two questions posed at the start of Sect. 3 should now be clear: it answers (b) by focusing on four agent roles that concern the use and production of artefacts and by characterising these roles in terms of

[15] See Hughes (2009) and Hughes' contribution to this volume for a way to make the connection between plans and means-end reasoning more explicit.

executing, constructing and communicating plans; this means that question (c) can be answered by introducing practical rationality and several consequent standards on plans, which indirectly apply to the agent roles. Individual actions are evaluated in the context of plans, where the prescriptive force may range from recommendation to requirement.

Finally, the plan analysis offers a way to study how prescriptive and descriptive statements are related within systems of technological knowledge. That there is such a relation seems beyond doubt: prescriptive statements that are not, somehow, related to true or justified descriptive statements are at best very risky recommendations. The general framework for discussing this relation is the role of justified beliefs in rational plans, that is, the standard of belief consistency discussed in Sect. 3. I end this section with, first, a few general remarks about this framework to sketch the answer to question (a) above and, second, a comment on Bunge's idea that technological rules are grounded in an explanation of their modus operandi.

Belief consistency offers a non-trivial constraint on rational plans, especially in our version, which requires justified beliefs. In the AI planning literature (on which Pollock 1995 provides a philosophical perspective), various ways are offered for explicitly constructing the 'belief base' of intentions and plans, often through DBI (desire-belief-intention) architectures. Any of these proposals is compatible with the plan analysis of prescriptive technological knowledge offered here. All avoid derivations of practical recommendations or requirements from descriptive statements, which were tentatively discussed by Bunge (1966) but seem highly problematic – both for general reasons, such as the risk of is-ought fallacies, and for the particular context sensitivities of the practical 'conclusions' involved.

There is another respect in which the present framework is more liberal and more realistic than Bunge's. Since it does not distinguish the prescriptive content of technological knowledge exclusively in terms of its underpinning, there is no need to insist on 'grounding' in law formulas. This creates room for extending the belief base to statements that express experience of previous successful use. As Bunge and, in more detail, Mokyr (2002) have argued, there may be a reason to replace such beliefs with ones that build upon a scientific understanding of the technology involved – but this reason is mainly to facilitate refinement of use practices or technological progress, not that evidence of success would not support knowledge of plans. Moreover, the belief base of users and makers can be minimal because they may rely on testimony for accepting the plans communicated to them by designers and manufacturers, respectively (see Houkes and Vermaas 2010, Ch.4). In implicitly assuming that everyone who knows technological rules or norms possesses a broad belief base for the corresponding plans, one ignores the social-epistemological aspects of technologies.

Finally, a remark on Bunge's emphasis on the explanation of the modus operandi of technologies. It can hardly be denied that professional designers, engineers in particular, typically possess knowledge about the physicochemical composition of artefacts and design artefacts on the basis of this knowledge. On the plan analysis, one way in which prescriptive and descriptive statements regarding artefacts are related is by means of a specific type of explanation for the function of an artefact

(Houkes 2006; De Ridder 2006). In such 'technological' explanations, which should be conceptually distinguished from the causal explanations offered by scientists, descriptions of the structure of an artefact are related to descriptions of the actions included in the use plan of the artefact to show that these actions can be rationally expected to lead to realisation of the goal state. That there may be these explanations does not mean, however, that designers and manufacturers should ground prescriptive statements in scientific knowledge in exactly this way, let alone that they are little more than 'applications' of this knowledge. The plan analysis does not distinguish between the strength of support given by the various sources (technological explanations, previous success and testimony), nor does there seem to be a systematic way – or a need – to do so.

5 Towards an Epistemology of Routines

As said at the outset, the original interest in specifying technology in epistemic terms sprung mainly from the perceived need to distinguish the efforts of engineers and engineering scientists from those of natural scientists. Leaving aside the major argumentative difficulties regarding this contrastive exercise, contemporary philosophical, historical or sociological research on technology manifests little need for it. No attempt will be made to revive it on the basis of the preceding analysis; the temptation will, for instance, be resisted to distinguish the prescriptive content of, say, engineering-science textbooks from that of textbooks in, say, electro- or hydrodynamics. For slightly different reasons, the results of the analysis will not be fed back into the debate regarding the 'design sciences', on which we touched briefly above and in which the notion of technological rules plays a supporting part.

Instead, a look is taken at another area of application: evolutionary approaches in economics and organisation theory. These approaches seek to incorporate processes of technological and organisational change in economic modelling. Often, this is part of an attempt to overturn or significantly modify neoclassical orthodoxy, with its emphasis on perfect rationality and equilibrium states. Here, I focus on one line of research within this variegated set of approaches, namely, that which analyses technological change through organisational 'routines'. This research, which found a paradigmatic expression in Nelson and Winter (1982), represents knowledge within complex organisations, such as industrial firms, in order to analyse, among other things, the emergence and diffusion of technologies and the resulting success or demise of firms. Thus, it focuses on technological knowledge in the wide sense, that is, on knowledge that is relevant to producing, marketing, using and otherwise being practically engaged with technology. This knowledge is represented on the level of organisations rather than individuals, by describing each (temporal stage of an)[16] organisation in terms of a determinate set of routines – where routines were

[16] Nelson and Winter and their successors model technological and organisational change as an evolutionary process, by looking at generations of populations of organisations and organisational routines, connected by the core Darwinian mechanisms of variation, selection and retention. Here, organisations may be allowed to change their routines over time, introducing the required divisions

originally characterised loosely as 'regular and predictable behavioral patterns' and then identified as the genetic basis for evolutionary change (Nelson and Winter 1982: 14).

This did, understandably, not go unchallenged in the literature. Internal debates within evolutionary economics and constructive criticisms by others have focussed on the cognitive basis for organisational routines, on their relation to individual (employee's) habits, on the analogy with genes and on the need for a genetic basis.[17] Although these debates are certainly relevant, they are sidestepped here – except for noting two things: first, the conclusion that 'progress on the routines-portion of the Nelson and Winter (1982) program has been slow' (Becker 2004: 663) and second, the absence of analyses of the epistemology of this 'routines-portion' to accompany analyses of its cognitive status and (neo-)Darwinist pedigree.

There are at least three reasons why such an epistemological analysis seems called for. First, the 'routines-portion' of Nelson-Winter models tends to be inappropriately one-sided. Taxonomical analyses of technological knowledge (e.g. Vincenti 1990; Faulkner 1994) have shown that organisations possess many different types of knowledge. Taking these together under one heading may be necessary because of modelling constraints (e.g. tractability or robustness). Such considerations are missing in the literature; instead, an appeal is made to 'know-how' or 'tacit knowledge' as an encompassing epistemological category. This appeal does not, however, stand up to critical scrutiny (Nightingale 2009), creating the need for a more fine-grained epistemology of technological knowledge in organisations. Such an epistemology must go beyond a taxonomy, although the mentioned taxonomies may be useful starting points.[18] Second, epistemological considerations add to the descriptive plausibility of the 'routines-portion'. Social and behavioural scientists have, in the last decades, become increasingly impressed by the limitations of human rationality. Perhaps in an attempt to sever any connection to neoclassical theory, some evolutionary economists appear to have driven this to extremes, describing all behaviour as 'routine-driven'. For an accurate description of the variation, selection and retention of technological knowledge, however, it can hardly (and need not!) be ignored that human agents are epistemically sensitive. Changes in routines, for instance, are not just the result of imitating others in a social context but may also result from rational deliberation. Likewise, ideas may be retained not just as a result of social or personal inertia but also because they are field-tested in Van Aken's (2004) sense. Adding deliberation, testing and other epistemological aspects of routines therefore adds to the palette of mechanisms to which models of technological and organisational change may appeal. Third and finally, adding epistemological aspects may remove

between generations on the level of temporal stages of organisations (say, fiscal years). Alternatively, organisations may not be allowed to change routines, dividing generations by moments at which new organisations may enter the market and/or unsuccessful ones disappear. The application of the plan analysis in this chapter is supposed to be neutral with respect to these two modelling approaches.

[17] See Becker (2004), Dosi et al. (2000), Hodgson (2004), Vromen (2006) and the references in these papers.

[18] Assuming, for the moment, that these taxonomies stand up to critical scrutiny.

the somewhat artificial separation between neoclassical and evolutionary approaches, by providing a normative twist to the latter.

In the remainder of this section, it is shown how the plan analysis of prescriptive technological knowledge may contribute to this research agenda. This can only be done in a qualitative, even programmatic way, which hopefully paves the way for a more detailed application.

First, the analysis differentiates various elements of technological knowledge that all contribute to productive organisational routines. Rather than black boxing everything between informational input (market signals, consumer polls) and organisational response (launch of new products, marketing campaigns) as 'behavioural patterns', the plan analysis incorporates various activities and divisions of labour within an organisation that connect input and output. Strategies for plan designing, product designing and manufacturing may thus be differentiated and possibly separated as loci of organisational change within Nelson-Winter models. Reconstructing designing and manufacturing in terms of plans, moreover, makes it possible to represent partial changes within these activities. In Sect. 3, the main elements of plans were listed, which also provide a rough measure of the similarity between plans. Thus, it may be modelled how (a prescriptive, plan-based) part of technological knowledge changes without assuming that knowledge gets completely abandoned or transferred wholesale. Furthermore, the interaction with consumers – and possibly outsourced production facilities – is reconstructed as the communication of plans as well as the transfer of services and artefacts. This allows a clearer conception of one of the prime phenomena that evolutionary economists seek to model, namely, technological change.[19]

Once it is acknowledged that organisational routines may be spelled out partly in terms of plans and procedural knowledge, additional mechanisms for explaining variation, selection and retention of these routines become available. In general, the communicability of plans between agents explains how routines come to be shared between individuals within an organisation. This, again, supplements the emphasis on 'tacit' knowledge, which may then be used to explain why knowledge depends on experienced professionals and is not easily shared within organisations or transferred between organisations. More in particular, the prescriptive force applying to plans and procedural knowledge may be used to explain – in part – the inertia of organisational practices. These need not be explained exclusively the result of individual habit (leading to the question, mentioned above, whether routines resemble habits on an organisational scale); instead, they may result from recommendations or requirements. Likewise, the social context of technological change may be studied in more detail through the consistency of plans with (beliefs about) societal regulations and adherence to socially accepted values. This may serve to strengthen retention mechanisms that apply to technological knowledge in organisations. Above, it was already indicated how adding deliberation and testing may increase the plausibility of the

[19] Cf. the distinction between product and service characteristics of technologies made by Saviotti and Metcalfe (1984) and developed further in Saviotti (1996).

'routine-portion'. In the context of the plan analysis, incorporating the standards of practical rationality would lead to an evaluation of procedural knowledge in terms of effectiveness and efficiency, providing a connection to rational choice theory and neoclassical economics. Furthermore, the various sources of evidence for plan effectiveness could be incorporated: testing, testimony (i.e. social-epistemological considerations) and 'grounding' in physical knowledge through technological explanations. This would undoubtedly complicate the dynamics of organisational routines. Yet models that, on the one hand, claim to represent actual cognitive processes within organisations more faithfully than neoclassical rivals but on the other hand oversimplify (or outright ignore) epistemological evaluation of these processes run the risk of being both normatively toothless and descriptively inadequate.

6 Conclusions

In this chapter, it was shown how analysing the design, use and production of technology in terms of plans identifies prescriptive content in technological knowledge. It was argued that the plan analysis provides a clarification, in several respects, for Mario Bunge's analysis of applied science in terms of technological rules. Like Bunge's analysis, the plan analysis focuses on prescriptions on actions. Unlike Bunge's analysis, it does not assume that such prescriptions may or even must be derived from scientific law formulas; it is limited to particular kinds of actions, and it analyses the standards applying to and the internal structure of prescriptive technological knowledge. Furthermore, a potential area of application was identified: analysing technological knowledge within organisations in terms of knowledge of plans or procedural knowledge, rather than exclusively in terms of routines and tacit knowledge, may enrich and substantiate models of technological change developed in evolutionary economics.

Acknowledgments Research for this chapter was made possible by the Netherlands Organisation for Scientific Research. Helpful comments on an earlier version were given by Marieke van Holland, Auke Pols and Krist Vaesen.

References

Anderson, J.R. 1981. *Cognitive Skills and Their Acquisition*. Hillsdale: Lawrence Erlbaum.
Audi, R. 1989. *Practical reasoning*. London: Routledge.
Becker, M. 2004. Organizational routines. *Industrial and Corporate Change* 13: 643–677.
Bratman, M. 1987. *Intentions, plans and practical reason*. Cambridge, MA: Harvard University Press).
Bunge, M. 1966. Technology as applied science. *Technology and Culture* 7: 329–349.
Cohen, M.D., R. Burkhart, G. Dosi, M. Egidi, L. Marengo, M. Warglien, and S. Winter. 1996. Routines and other recurring action patterns of organizations. *Industrial and Corporate Change* 5: 653–699.

De Ridder, J. 2006. Mechanistic artefact explanation. *Studies in History and Philosophy of Science* 37: 81–96.

Dosi, G., R.R. Nelson, and S.G. Winter. 2000. *The nature and dynamics of organizational capabilities*. Oxford: Oxford University Press.

Faulkner, W. 1994. Conceptualizing knowledge used in innovation. *Science Technology and Human Values* 19: 425–458.

Franssen, M. 2006. The normativity of artefacts. *Studies in History and Philosophy of Science* 37: 42–57.

Georgeff, M., and A.L. Lansky. 1987. Procedural knowledge. *Proceedings of the IEEE* 74(10): 1383–1398.

Groover, M.P. 2007. *Fundamentals of modern manufacturing*. New York: John Wiley.

Hansson, S.O. 2006. Category-specified value statements. *Synthese* 148: 425–432.

Hendricks, V.F., A. Jakobsen, and S.A. Pedersen. 2000. Identification of matrices in science and engineering. *Journal for General Philosophy of Science* 31: 277–305.

Hodgson, G.M. 2004. The mystery of the routine. *Revue Économique* 54: 355–384.

Houkes, W. 2006. Knowledge of artefact functions. *Studies in History and Philosophy of Science* 37: 102–113.

Houkes, W. 2009. The nature of technological knowledge. In *Handbook of philosophy of technology and engineering sciences*, ed. A.W.M. Meijers, 309–350. Amsterdam: Elsevier.

Houkes, W., and P.E. Vermaas. 2009. Contemporary engineering and the metaphysics of artifacts. *The Monist* 92: 403–419.

Houkes, W., and P.E. Vermaas. 2010. *Technical functions*. Dordrecht: Springer.

Houkes, W., P.E. Vermaas, C.H. Dorst, and M.J. de Vries. 2002. Design and Use as plans. *Design Studies* 23: 303–320.

Hughes, J. 2009. An artifact is to use. *Synthese* 168: 179–199.

Jackson, J.D. 1975. *Classical electrodynamics*, 2nd ed. New York: Wiley.

Kroes, P.A. 1992. On the role of design in engineering theories. In *Technological development and science in the industrial age*, ed. P.A. Kroes and M. Bakker, 69–98. Dordrecht: Kluwer.

Mitcham, C. 1978. Types of technology. *Research in Philosophy and Technology* 1: 229–294.

Moran, M.J., H.N. Shapiro, B.R. Munson, and D.P. DeWitt. 2003. *Introduction to thermal systems engineering*. New York: Wiley.

Mokyr, J. 2002. *The gifts of Athena*. Princeton: Princeton University Press.

Nelson, R.R., and S.G. Winter. 1982. *An evolutionary theory of economic change*. Cambridge, MA: Belknap Press.

Nightingale, P. 2009. Tacit Knowledge and Engineering Design. In *Handbook of philosophy of technology and engineering sciences*, ed. A.W.M. Meijers, 351–374. Amsterdam: Elsevier.

Niiniluoto, I. 1993. The aim and structure of applied research. *Erkenntnis* 38: 1–21.

Pollock, J.L. 1995. *Cognitive carpentry*. Cambridge, MA: MIT Press.

Saviotti, P.P. 1996. *Technological evolution, variety and the economy*. Cheltenham: Edward Elgar.

Saviotti, P.P., and J.S. Metcalfe. 1984. A theoretical approach to the construction of technological output indicators. *Research Policy* 13: 141–151.

Simon, H.A. 1981. *The sciences of the artificial*, 2nd ed. Cambridge, MA: The MIT Press.

Staudenmaier, J.M. 1985. *Technology's storytellers: Reweaving the human fabric*. Cambridge, MA: The MIT Press.

Van Aken, J.E. 2004. Management Research Based on the Paradigm of the Design Sciences. *Journal of Management Studies* 41: 219–246.

Van Aken, J.E. 2005. Management research as a design science. *British Journal of Management* 16: 19–36.

Vermaas, P.E., and W. Houkes. 2006. Use plans and artefact functions. In *Doing things with things*, ed. A. Costall and O. Dreier, 29–48. London: Ashgate.

Vincenti, W.G. 1990. *What engineers know and how they know it*. Baltimore: Johns Hopkins.

Von Wright, G.H. 1963. *Norm and action*. London: Routledge and Kegan Paul.

Vromen, J.J. 2006. Routines, genes and program-based behavior. *Journal of Evolutionary Economics* 16: 543–560.

Chapter 4
Beliefs, Acceptances and Technological Knowledge

Marc J. de Vries and Anthonie W.M. Meijers

1 Introduction: Can Technological Knowledge Be a Matter of Beliefs Only?

One of the four ways of conceptualizing technology that Carl Mitcham distinguished in his book *Thinking Through Technology* is technology as knowledge. His description of technology as knowledge showed that not much philosophical literature on the nature of technological knowledge was available at the time he wrote this book as far as the analytical approach in philosophy was concerned. This is changing now. There is an increasing interest in reflecting on the nature of technological knowledge among philosophers of technology who are oriented towards the analytical tradition, although the total amount of publications is still limited and there is still a lot to be desired (Houkes 2009). In our contribution to the Companion to the Philosophy of Technology Meijers 2009; (Meijers and de Vries 2009), we have argued for the following four characteristics of at least part of technological knowledge: collectivity (technical norms as content of technological knowledge require a community for their existence), context dependence, normativity and non-propositionality. At the Eindhoven University of Technology, in the Netherlands, the normative dimension in technological knowledge is the focus of philosophical research. We find this normative dimension in technological knowledge in various ways: knowledge of functions (i.e. of what an artefact ought to enable us to do) but also knowledge of technical norms and standards, of good practice, of requirements for designs. In his recent survey of literature on technological knowledge, Houkes suggests this topic

M.J. de Vries (✉) • A.W.M. Meijers
Eindhoven University of Technology, Eindhoven, the Netherlands
e-mail: m.j.d.vries@tue.nl; a.w.m.meijers@tue.nl

M.J. de Vries et al. (eds.), *Norms in Technology*, Philosophy of Engineering and Technology 9, DOI 10.1007/978-94-007-5243-6_4,

of normativity as one of the promising ways for the emancipation of technological knowledge as a separate epistemological domain (Houkes 2009). An interesting aspect of technological knowledge is that it poses problems for the account of knowledge that often serves as the starting point for debates in epistemology, namely, the 'justified true belief' account. There has been the well-known critique that this definition is limited to propositional knowledge, while much of technological knowledge is of a non-propositional nature. There are, for instance, technological skills that make any account of technological knowledge solely based on beliefs problematic. In addition there is the puzzle that effectiveness seems often more decisive for what engineers take to belong to their knowledge than the truth of their beliefs, particularly when it comes to normative knowledge, such as knowledge of norms.[1]

In this chapter we will not try to delineate exactly what is technological knowledge and what is not. For our purpose it suffices to take technological knowledge to be the knowledge that engineers use to design and make new artefacts and systems and the knowledge that the users of those artefacts and systems need to use them.

We take it for granted in this chapter that the 'justified true belief' account does not fully cover the nature of technological knowledge because of the non-propositional nature of certain types of technological knowledge (see also Chap. 2 by Meijers and Kroes in this volume). What is less obvious, though, is that not all propositional knowledge in technology is belief-based. This is an issue we will focus on in this chapter. We will argue for an acceptance-based account of certain types of technological knowledge.

A number of authors have argued that beliefs and acceptances are different mental states. Beliefs, for instance, are involuntary (I cannot decide whether I believe or not believe that p), but acceptances are voluntary (I can decide whether or not I accept that p). Another difference is that beliefs come in degrees (my belief that p may start as a fairly weak belief, and this belief may grow in the course of time), while acceptances are a matter of 'yes' or 'no' (either I accept that p or I do not accept it). Other differences are mentioned, but a careful study of what various authors have written shows that there is no general agreement on all the ways in which beliefs differ from acceptances. One of the differences concerns the context dependence or context independence of beliefs and acceptances. In Sect. 2, we will discuss the different views in more detail.

Why, then, is this discussion about belief and acceptance of interest to the debate about technological knowledge? That is because certain types of technological knowledge are good examples of propositional knowledge for which a belief-based account is inadequate and an acceptance-based account could be a promising alternative. The success of this effort, though, to a certain extent depends on what

[1] As far as prescriptions are involved in technological knowledge, one could furthermore wonder what it means for a prescription to be believed to be 'true'. Prescriptions can be adequate, but not true. Perhaps in normative realism this is possible, but even then the fact that prescriptions are hypothetical imperatives, conditioned on the desired change of state of affairs, makes it difficult to imagine how truth-values can be defined.

one takes to be an acceptance, and here the differences between different authors have to be examined to see what account fits best technological knowledge. But why search in the direction of acceptances? The whole issue of acceptance versus belief emerged in the context of practical reasoning. Beliefs were found incapable of accounting for various aspects of this type of reasoning, and the concept of acceptances was brought forward as an alternative. In technology the use of knowledge is intimately related to practical reasoning. Designers do a lot of practical reasoning to come up with designs and to justify them. For that reason, acceptances may be as relevant as beliefs to describe the nature of technological knowledge. A second reason for seeking in the direction of acceptances can be found in an article by Joseph Pitt, 'What Engineers Know', in which he suggested that technological knowledge, contrary to scientific knowledge, is task-dependent (Pitt 2001). If an airplane designer would be put to work in a car company and she/he would accept the same propositions about necessary safety margins, that would certainly be a cause of trouble in that company. As task-dependence implies context dependence, and a number of authors claimed context dependence to be a defining characteristic of acceptances, an acceptance-based approach of technological knowledge looks promising. This chapter is a first exploration of this idea.

2 Types of Acceptances

In 1992, Cohen published an essay on acceptances with the practical purpose of tackling a number of conceptual problems related to a belief-based nature of knowledge (Cohen 1992). In particular Cohen was not satisfied with the idea beliefs were the only mental state representing knowledge, given the fact that beliefs are involuntary (Hume is mentioned as the originator of this idea; Cohen, p. 20). Cohen saw the need to conceptualize a different mental state with a more active nature: acceptances. Beliefs, according to Cohen, are dispositions to feel, while acceptances are policies for reasoning (Cohen, p. 5). From this he derives that acceptances are voluntary and beliefs are not (Cohen, p. 23). Knowing that p then entails either the acceptance that p or the belief that p, depending on the context (Cohen, p. 99). Cohen went on to seek other differences between beliefs and acceptances and claimed that beliefs are supported by evidence, whereas acceptances can also (or instead) be supported by prudence (Cohen, p. 12). We may want to accept that it will rain this evening even though we have no evidence for that, but because we want to be on the safe side and take our umbrella with us. That would not hold as a reason to make us believe that it will rain this evening. Another distinction is that acceptances, unlike beliefs, are relative to context, and not a matter of degrees (Cohen, p. 16). Finally, acceptances can be said to be 'subjectively closed under deducibility', that is, anyone accepting p1, p2, …, pn, and the deducibility of q from that conjunction also accepts q ('objectively closed' would mean that any q deducible from the conjunction would have to be accepted, irrespective of whether or not the person would recognize its deducibility; Cohen, p. 29). Beliefs, on the other hand, do not carry such commitments (Cohen,

p. 31[2]). From that Cohen derived that a belief that p may be combinable with a simultaneously held belief that non-p. The same can be said of acceptances, but this would count as a form of irrationality (Cohen, p. 35).

In the same year 1992, Bratman published a paper in which he used the concept of acceptances with a different purpose, namely, to account for the role of knowledge in practical reasoning (Bratman 1992). Bratman had been accused by Cohen of being unaware of the need to distinguish between beliefs and acceptances in a 1987 publication, and in the 1992 paper Bratman takes up the challenge posed by Cohen. For Bratman, beliefs are context-independent, shaped by evidence, not subject to voluntary control, and subject to an ideal of integration (or 'subjective deducibility' as Cohen called it; Bratman, p. 4). At this last point he differs from Cohen because he sees beliefs and acceptances both as inner states rather than one as a disposition and the other as a mental act. Acceptances differ from beliefs in that they are context-dependent. Bratman does not discuss other differences in this 1992 article.

In 1998 Pascal Engel points out the need to add a second description of acceptances, different from the ones Cohen and Bratman used (which he indicated as acceptance$_1$, assent or 'holding true'). For Cohen and Bratman acceptance is the assent to a proposition and differs from the corresponding belief only in that it is voluntary, context-dependent and not a matter of degrees. Engel sees a need to define a second version of acceptance[3]: acceptance$_2$ or 'holding as true'. The difference between the two is that acceptance$_1$ is prompted by a corresponding belief, whereas acceptance$_2$ is not (Engel 1998, p. 147). Consequently, acceptance$_2$ is not necessarily aimed at truth (as belief is) but at utility or success and does not fall under the commitment of subjective deduction (or integration, as Engel calls this). In Table 4.1 Engel's characteristics of his acceptance$_2$ and the differences with belief have been listed.

In a 2000 article in Philosophical Explorations, Raimo Tuomela also uses the term 'holding true' but not in exactly the same sense as Engel used it (Tuomela 2000). The difference between Tuomela's 'holding true' and Engel's use of the same term for acceptance was noted by Engel in a later article (published in 2004), in which he indicated that Cohen's use of the term 'holding true' prompted for the identification of a third type of acceptance: acceptance$_3$ or 'accepting as true' (Engel 2004). The difference between belief and acceptance$_3$ is in a deliberate decision to accept as true what previously was held as a belief.[4] That decision causes a transition

[2] It should be noted that the debate about the closure of beliefs is still unresolved in contemporary epistemology. There is at least the problem that in the case of contradictions among p's any q can be deduced. This problem is even more pressing for acceptances than for beliefs, as acceptances are said not to be aimed at truth.

[3] At least, this is held in most publications on this issue. It may be correct that being truth-aimed may not have the explanatory value that was often ascribed to it (to explain the facts that correct beliefs are true beliefs, that rational beliefs are supported by evidence and that we cannot form beliefs at will), as Owens argues, but that does not mean being truth-aimed is not a property of beliefs.

[4] This type of acceptance had also been identified by Keith Lehrer previously (1999).

Table 4.1 The various types of acceptances, as compared to belief

	Belief (proper)	$Acceptance_1$ (holding true)	$Acceptance_2$ (holding as true)	$Acceptance_3$ (accepting as true)
Voluntary?	No	Yes	Yes	Yes
Aimed at truth?	Yes	Yes	No	Yes
Shaped by evidence?	Yes	No	No	Yes
Subject to integration?	Yes/no[a]	Yes	No	Yes
Comes in degrees?	Yes	No	No	No
Context-dependent?	No	Yes	Yes	No

[a]'Yes' according to Bratman and Engel, 'No' according to Cohen

from the involuntariness of the belief to the voluntariness of the decision to accept as true, and also it makes a coming in degrees of the belief end in an 'all-or-nothing' decision to accept (or reject) what was previously believed. As the belief had been shaped by evidence, $acceptance_3$ would also be, and this makes it different from both $acceptance_1$ and $acceptance_2$. In Table 4.1 the various descriptions of acceptance according to Engel are presented.

This table indicates that there is a whole spectrum of acceptances. One may wonder whether there are even more than the three types that are listed in the table.

3 Types of Technological Knowledge

Given these differences between belief and acceptance, it is worthwhile to explore how they possibly relate to various types of technological knowledge distinguished in the literature. We will use here the taxonomy of technological knowledge that de Vries developed earlier (de Vries 2003). His taxonomy is based on a conceptualization of technical artefacts as objects having a dual nature: a physical nature and a functional nature.[5] From this dual nature, three types of artefact-related technological knowledge can be derived: knowledge of the physical nature, knowledge of the functional nature and knowledge of the relationship between physical and functional nature.[6] de Vries has also argued that there is a normative dimension in some of these knowledge types that we do not find in scientific knowledge.[7] We will

[5] This conceptualization was the basis of the "Dual Nature of Technical Artefacts" research project that was carried out at Delft University of Technology; see www.dualnature.tudelft.nl.

[6] In addition to these three types, de Vries added a fourth: knowledge of processes (i.e. knowledge of the sequence of actions that makes the artefact function possible). In this chapter we will not discuss that type. We have to add here that all these knowledge types only refer to propositional knowledge. Knowing how is not included in this taxonomy.

[7] This normative component is the key issue in the "Norms in Knowledge" research project that is carried out at Eindhoven University of Technology.

take this taxonomy as our starting point for exploring the belief or acceptance base of technological knowledge.

The first type of artefact-related knowledge concerns knowledge of the physical nature of the artefact: A person S knows that artefact A has physical property p. Although it can hardly be expected that this type of knowledge is acceptance-based (this would almost immediately imply that all scientific knowledge is acceptance-based), we will briefly indicate why the characteristics of beliefs rather than acceptances apply to this type of knowledge. In this proposition, p can refer to artefact properties such as size, weight, materials it is made of and solvability. Knowledge of such properties is often based on the perception of the artefact. The knowledge that person S forms about the artefact is then involuntary, as is usual for beliefs that are formed through perception. The same would hold if S would form such knowledge through memory, testimony or reasoning. This knowledge can be said to be truth-aimed. S wants to acquire knowledge that corresponds to the real physical nature of the artefact. That holds both for the user for whom this knowledge can be the basis for ascribing functions to the artefact, as well as for the designer who has chosen the physical properties in order to allow such functions to be fulfilled. Such knowledge is shaped by evidence, of whatever kind (S's own observation or otherwise). This knowledge is subject to integration. S cannot believe at the same time that A is made of unpainted gold and is blue and extremely light in weight. Such knowledge allows for degrees. At first S may not be entirely sure that A is made of gold, but by more careful inspection and by reasoning, she/he can become more and more convinced that it is the case. There seems to be no context dependence in this type of knowledge. S will claim to know that A is shaped round, independent of the situation in which she/he finds the artefact. Taken together, these characteristics of knowledge of the physical nature suggest that it is belief-based. As with all beliefs, however, beliefs about the physical nature of A can be transformed by S into acceptance$_3$ by a deliberate decision to accept as true what was believed (see Table 4.1). As stated before, this conclusion was to be expected and the arguments were just presented for the sake of completeness of the defence of our claim that part of technological knowledge is belief-based. Natural science knowledge is an integral part of technological knowledge.

The second type of knowledge concerns knowledge of the function of an artefact. This knowledge is of the following kind: S knows what an artefact is for, what its function is. He knows that carrying out action ACT with artefact A *should* result in a change of state of affairs $\varphi_i \rightarrow \varphi_{i+1}$. This knowledge is not involuntary. Having seen A, S has different options for ascribing a function to A.[8] Perhaps she/he has used A before or saw other people use it and thus has developed (involuntary) beliefs about its particular function. But this still leaves it open to S which function to ascribe to A. This is different from knowledge of the physical nature

[8] As we consider acceptances here as a basis for knowledge, this function ascription cannot just be a random guess. Some sort of epistemic accomplishment has to be involved. This can be the experience someone has had with similar devices.

of A: having seen a table, I cannot but believe that it is made of wood. But knowing that a particular table is for sitting at, rather than for standing on, is based on a voluntary acceptance that that is what that table is for, what its very function is. There is no necessity here. It could also have been a work bench or a piece of art. This type of knowledge is shaped by considerations of effectiveness or success rather than by truth.[9]

When I claim to know that the screwdriver I have in my hands can be ascribed to the function of opening a tin can, is this type of knowledge only shaped by evidence? For a belief this is necessary, but here evidence may not be the only consideration, as prudence or trust or other practical considerations can shape this type of knowledge as well. I may have no evidence available that the weapon I have in my hand will really be effective in killing the grizzly bear in front of me as I have never heard that someone else has used it for killing a grizzly bear, but for prudential reasons it makes sense to accept it as a killing device and use it rather than dropping it and fighting with one's bare hands. In this case it may seem rather far-fetched to speak of knowledge. In engineering, though, there are similar cases in which it is not uncommon to speak of knowledge. For instance, when an engineer trusts that a certain drilling machine will be able to cut through a certain material that she/he has not worked with before, and thus accepts as the function of a certain drilling machine that it will cut through that material, that commonly counts as part of his/her knowledge. It shows that she/he is able to make educated guesses about which artefact to choose and use from a variety of possibilities. Wright has argued that trust without evidence can be quite acceptable as an epistemic entitlement for knowledge (Wright 2003, 204). But then this knowledge must be acceptance-based, as beliefs cannot be rationalized by practical considerations only, as Kelly has argued (Kelly 2002, 177).

Is knowledge of functions subject to integration? Prima facie that seems to be the case. S cannot at the same time be convinced that A can be used for covering an electric wire for safety reasons and at the same time be used to connect a bulb to a battery. But we have to be careful here. The reason that S cannot believe those two things simultaneously is not caused by knowledge of the functional nature, but by reasoning back to the physical nature (and concluding that A cannot at the same time isolate and conduct electricity). Had S not reasoned like that, she/he may have had no problem in holding the idea that both functions could be realized with the same artefact A. So the users' knowledge only becomes subject to integration when

[9] In addition to these function-related beliefs aimed at effectiveness, one can develop associated beliefs that are truth-aimed; for instance, I can read an article that reports the artefacts' effectiveness for a certain function having been tested. One can also question if correspondence with reality is at all involved in this type of knowledge. Even if ACT is carried out and A malfunctions, S can still claim that she/he knows A *should have* enabled $\varphi_i \rightarrow \varphi_{i+1}$. This would not be the case in a claim like: "this stone should drop when I let it go". In that case I would abandon the claim in case the stone would not drop, because reality is the decisive criterion for such a claim. But when a car fails to start, WE will still uphold the claim that as a car it ought to enable me to get from A to B.

knowledge of the relationship between functional and physical nature comes in. It is conceivable that users ascribe conflicting functions to artefacts when they lack knowledge of that relationship.[10]

Does knowledge of functions come in degrees? Not necessarily so. S, based on many years of experience, may claim that she/he fully knows that realizing $\varphi i \rightarrow \varphi i+1$ can be a function of A, in spite of whatever the manual says or what other people say or what a first tryout has shown. In other cases, though, S may feel more and more confirmed by the idea that A has that function, and in that case the knowledge seems to be a matter of degrees and not an all-or-nothing matter. In the case of the grizzly bear attack, my acceptance of the idea that my weapon has the function of a grizzly bear killer may well grow as the bear is getting closer and closer to me. But a shift in the opposite direction is also possible. Clearly there are characteristics involved in the knowledge of the functions that suggest that this type of knowledge is not based on beliefs but on the acceptance of a view in which practical considerations play a role.

Is knowledge of functions context-dependent? That seems to be the case indeed, as we have seen before. Depending on the situation and the needs and desires of S related to that situation, S may show knowledge of different functions that can be ascribed to A. A person may attribute the function of protection against rain in a situation of a drizzle, but she/he will deny a protective function to the same device in the situation of a tropical storm as protecting against rain and against tropical storms can be considered to be two different functions. This could not be the case for beliefs about the physical nature of the artefact. Our analysis so far seems to suggest the following characteristics of functional knowledge: to a certain extent voluntary, based on considerations of practical success instead of truth, not only shaped by evidence, subject to integration (but need not be), does not always come in degrees, but is always context-dependent. This matches roughly with the characteristics of acceptance$_2$ in Table 4.1.

What can we say about knowledge of the relationship between physical and functional nature? Here we have to distinguish between users' knowledge and designers' knowledge. The users' version of this type of knowledge[11] is of the following kind: S knows that A's physical property p (or a combination of properties pi) makes it suitable for carrying out with A the action ACT that results in the change in state of affairs $\varphi_i \rightarrow \varphi_{i+1}$. The designers' version of this type of knowledge is

[10] Here, again, we have to ask: can we speak of knowledge? This can, indeed be the case. If the user has experienced that a diode can be used to block an electric current in a circuit but the same diode can be used to conduct that current, then the ascription of those two conflicting functions can be called knowledge. It is only by reasoning back to the relation between the physical properties of the diode and these two functions that the user realizes that the functions are not really in conflict. But as long as this relationship knowledge is absent, it may well count as an epistemic accomplishment, and therefore as knowledge, that a user accepts the two (seemingly) conflicting functions of the same diode.

[11] Of course designers can have this version too, but we use the term 'users' version' to indicate that for users it is more prominent than for designers. We will use the term 'designers' version' in a similar way.

different: in order to let action ACT with A result in $\varphi_i \rightarrow \varphi_{i+1}$, A should have physical property p (or a combination of physical properties pi). These two versions differ considerably. The user starts with the physical nature of the artefact at hand and from that seeks possible functions. The designer starts with desired functions and from that she/he seeks a suitable physical nature (properties). Yet, we take these as two species of one type because both relate the physical to the functional nature.

Let us first examine the users' version. This type of knowledge is descriptive ('I know that A is suitable for ACT'), just like knowledge of the physical nature is. Observing the artefact is usually a crucial step in obtaining this type of knowledge and this implies involuntariness of the resulting knowledge. When I see the artefact and recognize it as something heavy, it is not a matter of free decision to obtain the idea that A could be used to strike a forceful blow. The idea that A's physical properties make it suitable for realizing $\varphi_i \rightarrow \varphi_{i+1}$ therefore seems to be involuntary (although this is not a unique relation, there may be many functional properties realized by very the same physical basis). Also it is truth-aimed, and this truth can be tested experimentally if desired by S. Here is a difference with knowledge of an artefact's functional nature: in case A appears not to be suitable for realizing the envisioned function, it is rational to drop the idea of A's suitability for realizing $\varphi_i \rightarrow \varphi_{i+1}$, while in the case of a malfunctioning device, the claim that it ought to function can be kept. So this type of knowledge is shaped by evidence rather than by prudence. Also this kind of knowledge is subject to integration, as we remarked earlier on. This knowledge can come in degrees, as further reflection on the relationship between physical properties and possible functions ought to enhance the conviction of A's suitability for a certain function in case of a rational user. Finally the knowledge that a certain physical property makes A suitable for a certain function is not context-dependent. The context dependence is in the ascription of the function, but not in the judgement about suitability. All together, this type of knowledge seems to be belief-based.

The designers' version of this type of knowledge, though, has the normative dimension that we also encountered in functional knowledge. She has knowledge of an abstract entity: a design with functional properties that has yet to be made physically (though she knows its physical properties). There is a certain voluntariness or looseness in this type of knowledge: in a different situation the designer could opt for a different property to create a fit with the desired function. That means there is also context dependence. Effectiveness or practical considerations may be more important than truth for designers. Furthermore, their knowledge can be shaped by prudence and not only by evidence (in case of safety considerations for example). Finally, this type of knowledge has a degree rather than an all-or-nothing character. A designer may be more or less convinced of the idea that the designed artefact will actually deliver its function. Comparing these characteristics with Table 4.1, we seem to have acceptance$_2$ as the basis for this type of knowledge. So there is an asymmetry in the knowledge that connects from physical to functional nature and the knowledge that connects from functional to physical nature, which is most likely caused by the presence of normativity in the latter and its absence in the former type of knowledge. This underlines the importance of taking into account normativity in characterizing technological knowledge.

4 Conclusions

Our analysis of technological knowledge strongly suggests that not all propositional knowledge in technology is belief-based. There are at least two types of technological knowledge that seem to be acceptance$_2$-based. Knowledge of the physical nature of an artefact is both belief-based and acceptance$_3$-based.

The conclusion raises the question: why was there no acceptance$_1$ found in the analysis? Let us examine once more the differences between beliefs and the various types of acceptances (see Table 4.1). The table shows that belief and acceptance$_2$ are the ends of a spectrum as they differ in each of the characteristics that are listed in the table. Acceptance$_3$ is closest to belief. It differs from belief in that it is voluntary rather than involuntary and comes in an 'all-or-nothing' way rather than in degrees. Acceptance$_3$ is the deliberate acceptance of what was until then a belief. One step further away from belief is acceptance$_1$, where we find two more differences with belief: it is not only shaped by evidence (but, e.g. also by prudence) and it is context-dependent. But it is still aimed at truth and subject to integration, as belief is. Why then is this combination absent in the various types of technological knowledge that we have examined? An obvious reason is that the types of knowledge we have evaluated do not exhaust all propositional knowledge in technology. We have only used knowledge related to the dual nature of technical artefacts. We left out knowledge of technical norms, for example. This knowledge can take the following shape: 'I know that in this situation technical norm N applies.' Could that be an example of acceptance$_1$-based knowledge? Pascal Engel describes acceptance$_1$ as holding true a sentence, not a proposition. One can believe a sentence to be true without understanding its content (for instance, when the sentence contains words that are unknown to the believer). This would mean that someone would hold a sentence about an artefact to be true without understanding what the sentence means. In principle, that could hold for knowledge of norms (for instance, safety norms, ISO standards). Norms are sometimes used without a full comprehension of their origin or nature. But these cases seem rare and do not add much to our analysis.

We conclude from our analysis of different types of (artefact-related) technological knowledge that at least some types are not belief-based but are acceptance-based.[12]

[12] The analysis of the various differences between belief and acceptance is also relevant for the analysis of technological knowledge in the following way. Both notions are central in the debate about collective intentionality (see, e.g. Meijers 2002). Acceptances in that context are an attractive alternative to beliefs, as collective acceptances do not seem to require difficult constructs such as a collective mind (which to some extent seems to be a prerequisite for collective beliefs, Wray 2001). It would be interesting to see whether the notion of collective acceptance could be used to enhance our understanding of technological knowledge. A lot of engineering knowledge is shared knowledge that functions in a community of engineers. That holds in particular for prescriptive knowledge, the basis of which often is a collective decision to accept certain norm or standard. Furthermore, Marcel Scheele convincingly argued that function ascription to artefacts has a social dimension. Hence, the knowledge of artefact functions also has a certain collectiveness.

References

Bratman, M.E. 1992. Practical reasoning and acceptance in a context. *Mind* 101: 1–15.
Cohen, L.J. 1992. *An Essay on Belief and Acceptance*. Oxford: Oxford University Press.
de Vries, M.J. 2003. The nature of technological knowledge: Extending empirically informed studies into what engineers know. *Techné* 6(3): 1–21.
Engel, P. 1998. Believing, holding true, and accepting. *Philosophical Explorations* 1: 140–151.
Engel, P. 2004. Is acceptance independent from belief?. Unpublished conference paper. http://jeannicod.ccsd.cnrs.fr/docs/00/05/33/49/HTML/index.html. Accessed 1 May 2010.
Houkes, W. 2009. The nature of technological knowledge. In (Handbook of the Philosophy of Science), ed. A. Meijers, 9: 309–350. Amsterdam: North Holland.
Kelly, Th. 2002. The rationality of belief and some other propositional attitudes. *Philosophical Studies* 110: 163–196.
Lehrer, K. 1999. Justification, coherence and knowledge. *Erkenntnis* 50: 243–258.
Meijers, A.W.M. 2002. Collective agents and cognitive attitudes. *Protosociology* 16: 70–85.
Meijers, A.W.M. (ed.). 2009. Philosophy of technology and engineering sciences, 1st edn. (Handbook philosophy of science), vol. 9, 1453 pp. Amsterdam: North Holland.
Meijers, A.W.M., and M.J. de Vries. 2009. Technological knowledge. In *A companion to the philosophy of technology*, ed. J.K. Berg Olson, S.A. Pedersen, and V.F. Hendricks, 70–74. Chichester: Wiley-Blackwell.
Pitt, J.C. 2001. What engineers know. *Techné* 5(3): 17–30.
Tuomela, R. 2000. Belief versus acceptance. *Philosophical Explorations* 2: 122–137.
Wray, K.B. 2001. Collective beliefs and acceptance. *Synthese* 129: 319–333.
Wright, C. 2003. On epistemic entitlement. Part I. Warrant for nothing (and foundations for free)? *Aristotelian Society Supplementary* LXXVIII: 167–212.

Chapter 5
Policy Objectives and the Functions of Transport Systems

Holger Rosencrantz

1 Introduction

The general purpose of this chapter is to identify lessons learned from policy objectives and to relate these lessons to key areas from modern philosophical literature. This chapter more specifically focuses on the Swedish transport policy objectives, which provide unusually clear illustrations of the problems involved in managing large and complex policy areas. The analysis aspires to highlight these problems. The Swedish transport policy objectives can be compared with international transport policy objectives as well as with objectives from other policy areas, suggesting that this specific case has certain features of general interest.

This chapter uses the terminology "goals" and "objectives" interchangeably. The policy literature (as well as the philosophical literature) sometimes uses the term "policy goal" rather than the often more adequate term "policy objective" (Rosencrantz 2009). The reason may well be that some languages lack the English distinction between "goal" and "objective" (The difference is that goals are either achieved or not, whereas objectives can be achieved to different degrees). The term "objective" also happens to be homonymous to an adjective of philosophical importance. In this chapter, the term "goal" is occasionally used when actually referring to objectives. It is of course desirable to strive towards terminological stringency. However, the two terms are related in the sense that an objective can be seen as a set of goals to achieve gradual achievement of the objective. Furthermore, many policy objectives are indeed visionary.

Based on observations from the specific case of Swedish transport policy goals, three general arguments and their normative implications are developed in this chapter. First, goals are subject to evaluation and updating. This observation has

H. Rosencrantz (✉)
Royal Institute of Technology, Stockholm, Sweden
e-mail: hrz@kth.se

M.J. de Vries et al. (eds.), *Norms in Technology*, Philosophy of Engineering and Technology 9, DOI 10.1007/978-94-007-5243-6_5,

fundamental implications for the relation between rationality and action. Second, there is a trade-off between precision and flexibility in the act of goal setting. These two properties are related to the two requirements of coordination and adaptation, respectively. Third, different kinds of goals require different approaches to implementation. A goal that reflects the purpose or function of a policy area is qualitatively different from a goal that reflects minimization of negative side effects of policy decisions.

The outline of this chapter is as follows: Sect. 2 provides a presentation of the Swedish transport policy objectives as well as a theoretical background. Based on this background, Sect. 3 presents the three main arguments as provided above. In Sect. 4, the lessons learned are related to selected topics from modern philosophical literature.

2 Background and Observations

2.1 Swedish Transport Policy Objectives

In Sweden, an overall objective to guide the design of its transport system was adopted by Parliament in 1998.[1] This objective is to provide a transport system that is both economically efficient and sustainable in the long term. This formulation followed a long tradition following the objective to provide "adequate transport at the lowest possible cost",[2] as formulated in similar Parliamentary decisions from the 1960s and onwards. In the decision from 1998, five sub-objectives were adopted to "specify" the overall objective: an accessible transport system, a high standard of transport quality, safe traffic, a good environment, and positive regional development. In a supplementary decision from 2001, the objective of a transport policy providing equal opportunities for men and women was added as a sixth sub-objective.

The Swedish transport policy objectives reflect current political priorities and are intended to guide regional and local authorities as well as central administration of the transport system. In 2009, the Swedish transport policy objectives were revised. The overall objective to provide an economically efficient and sustainable transport system has remained unchanged, but the structure of the sub-objectives was changed. Instead of six separate sub-objectives, there is now a "functional objective" and an "impact objective"[3]:

[1] The Swedish transport policy objectives and its predecessors are presented in the governmental bills prop. 1963:191, prop. 1978/79:99, prop. 1987/88:50, prop. 1996/97:137, 1997/98:56, prop. 2001/02:20, and prop. 2008/09:93.

[2] In Swedish: tillfredsställande transportförsörjning till lägsta möjliga kostnader.

[3] "Impact objective" is the government's own translation of the Swedish compound term *hänsynsmål*, literally meaning "heed objective".

- *Functional objective* (accessibility): The design, function, and use of the transport system will contribute to provide everyone with basic accessibility of good quality and functionality and to development capacity throughout the country. The transport system will be gender equal, meeting the transport needs of both women and men equally.
- *Impact objective* (health, safety, and the environment): The design, function, and use of the transport system will be adapted to eliminate fatal and serious accidents. It will also contribute to the achievement of the environmental quality objectives and better health conditions.

Transport quality is now part of the functional objective, under the heading "accessibility", which also covers the objectives of equal opportunities and regional development. The environmental objective (which refers to a separate set of objectives in the Swedish environmental policy) and the safety goal are part of the impact objective. The simplified structure is intended to facilitate implementation of the objectives and trade-offs between conflicting objectives.

For the sake of generality, it is useful to compare the Swedish transport policy objectives to some non-Swedish transport policy objectives.[4] It is common to formulate national transport policy as a set of objectives reflecting current political priorities. In Finland, for example, transport policy objectives concerning financing, regional success, environmental impact, and traffic safety have been formulated. The UK Department for Transport has five strategic objectives, including economic growth, greenhouse gases reduction, safety, equal opportunities, and quality of life. Similar transport policy goal structures can be found in, for example, Norway, the Netherlands, Australia, and South Africa. A transnational example is the goal of the European Union's transport policy, which involves sustainability, social inclusion, and promotion of Europe's competitive advantage.

In Sweden, the transport policy objectives are operationalized by a number of targets. For example, the safety goal is operationalized by the target to halve the number of road fatalities and reduce the number of serious injuries by a quarter during the period 2007–2020. The environmental objective is operationalized by, among other targets, the target to make Sweden's vehicle fleet independent of fossil fuels by 2030. Further details on the objectives are provided in strategy documents, which are adopted at local and organizational level.

The safety goal, which in 2009 became a part of the impact objective, deserves special mention. For various reasons, this goal has been the subject of much attention and criticism (Rosencrantz et al. 2007). The so-called Vision Zero, which was

[4] An overview of the transport policies of selected European nations is provided in report 2008:2 of the Swedish Institute for Communication Analysis. The Directions for Traffic Policy and Development published by the Finnish Ministry of Communications, the 2009 Autumn Performance Report of the UK Department for Transport, the Australian National Transport Policy Framework, and the South African White Paper on National Transport Policy are examples of publications providing comprehensive overviews of individual nations' transport policies. An overview of the transport policy of the European Commission can be found on, for example, the website of the DG for Energy and Transport.

adopted by Parliament in October 1997, states that the long-term goal for Swedish road safety policy is that nobody should be killed or seriously injured (as defined by the need for more than 3 weeks of hospitalization) as a consequence of accidents in the transport system. While this level of ambition itself is notable, a key feature of the safety goal is the shift of responsibility from the user to the designer; rather than seeing fatal traffic accidents as a tragic but necessary part of the transport system and as largely caused by the human factor, Vision Zero states that the transport system should be designed such that no fatalities or serious injuries occur (Johansson 2009). As a corollary, the transport system is to become more forgiving; a mistake from the user's side (inappropriate use) is not to lead to a "death sentence". In line with this thinking, Vision Zero states that unprotected users of the transport system, such as pedestrians and bicyclists, are to be protected. Vision Zero has been characterized as a systemic or holistic approach to road safety, in contrast to approaches that focus on human error as the main cause of serious accidents (Larsson et al. 2010). Finally, Vision Zero explicitly focuses on personal damage rather than property damage; while deaths and serious injuries are not acceptable, accidents leading to material damage only may still be regarded as acceptable.

Internationally, transport safety is an area in which ambitious goals and quantified targets are particularly prominent. In 2000, the Norwegian Parliament adopted a vision of zero killed and seriously injured. The Danish government has formulated its vision in the slogan "every accident is one too many". The Road Safety Council of Hong Kong is committed to the goal "zero accidents on the road". The European Commission's third action programme for road safety, which included a target to halve the number of road deaths between 2003 and 2010, has been recognized as a strong catalyst of efforts made by member states to improve road safety and triggered the process to propose a goal to halve the number of road deaths during the period 2011–2020. Traffic accidents pose a great challenge to developing countries, with fast growing vehicle fleets combined with seriously underdeveloped road safety; as road traffic accidents represent a substantial share of the global burden of disease, a level of harm said to be "unacceptable" by the WHO and the World Bank in their first major joint report on the subject, 2011–2020 has been globally proclaimed the "decade of action for road safety" by the UN.

For the sake of generality, a short comparison should also be made between transport and other policy areas. Policy objectives are also important in non-transport areas. Notably, environmental policy in Sweden and in other countries is also extensively managed by national objectives reflecting political priorities. National objectives for workplace safety are also prominent, and many comparisons can be made to road safety; for example, the Australian Occupational Health and Safety strategy builds on a "national vision" of zero fatalities and "national targets" of quantified reductions of fatality rates. In 2008, the Swedish government also adopted a "vision zero" for suicide.

Three features of the Swedish transport policy objectives should be highlighted. First, the current policy objectives have succeeded other objectives. While there is some continuity in the prevalence of economic efficiency as an overall goal, the reformulation of the objectives has reflected shifts in political priorities, organizational restructuring, and a perceived need for simplification. Second, the objectives and targets differ in how specific they are. While the overall goal of economic efficiency and sustainability provides relatively little information on which specific actions to prioritize, the targets of a vehicle fleet independent of fossil fuels by 2030 and a reduction of road fatalities by 50% during 2007–2020 state specific results to be achieved during a specific period of time. Third, the distinction between functional objective and impact objective that is present in the last version of the Swedish transport policy goals suggests that there is a qualitative difference between these two objectives. While accessibility is the function or purpose of the transport system, decision-makers must also ensure that transportation does not lead to negative impacts in terms of health, safety, and the environment. Before exploring the general normative implications of these three observations, some theoretical background of relevance will be provided.

2.2 Conceptions of Objectives and Rationality

Goals and objectives are important concepts in both philosophy and in policy sciences. While it is beyond the scope of this chapter to provide a comprehensive account of how goals and objectives are perceived in these traditions or in other relevant traditions such as artificial intelligence and organizational management, it is useful to provide some illustrations of how these concepts have been related to concepts of rationality. As rationality is a fundamental normative concept and as such important in establishing lessons learned from the case of the Swedish transport policy goals, it is worthwhile to identify different interpretations of the goal-rationality relation. This section also serves to identify theoretical background material to the observations highlighted in the previous section.

Traditional concepts of rationality typically stress the role of reason, but reason has been given different roles in different interpretations of rationality. A classical distinction is between the Platonic image of reason as a force that controls the passions, as in the chariot allegory, and the Humean image of reason as a slave of the passions. Analogous to this distinction from philosophy, the policy literature features the two topics of goal setting and goal achieving.

Goals and objectives are important elements within both descriptive and normative policy studies, and it has been recognized that goals and objectives are crucial both in evaluating public policy and in implementing organizational policy (Cingranelli et al. 1980; Rawson 1980; Nagel 1984). Conceptually, policy evaluation is a matter of identifying goals, measuring relevant attributes, and controlling for other possible sources of variation; among the complications noted in this process are that policy objectives tend to evolve during the course of implementation and that the setting of policy objectives affects the implementation itself, in other words, that goal setting and implementation are not two separate activities that could be viewed in isolation from each other. In policy theory, there is a linear concept of rationality which depicts policy-making as moving through distinct steps of policy agenda setting, formulation, decision-making, implementation, and evaluation (Parsons 1995; Sabatier 1999). Critics of this concept of rationality have in different ways stressed the interrelation between goal setting and implementation and that the starting point of rationality is not a predefined problem or objective; policy theorists have compared this to John Dewey's philosophy, in which the character and value of means and ends are reciprocally determined (Kuruvilla and Dorstewitz 2010).

In philosophy, practical rationality has been interpreted both as a relation between actions and objectives and as a property of objectives themselves. The former has been associated with what is sometimes called instrumental accounts of rationality. Instrumental accounts of rationality deny that it is at all meaningful to refer to objectives as rational or irrational. Instead, it is stressed, actions are said to be rational if they efficiently advance the achievement of the agent's objectives. Noninstrumental accounts of rationality typically stress that both interpretations (of rationality as a relation between actions and objectives and as a property of objectives themselves) are meaningful. Variations of the instrumental account have been ascribed to Aristotle, Hume, Russell, Allais, and Simon, while some writers such as Sen and Nozick have proposed theories of rationality beyond the instrumental account. Where goal-setting rationality has been a topic, it has been argued that a goal needs to coordinate actions and induce its own achievement if it is to be assessed as rational (Edvardsson and Hansson 2003; Rosencrantz 2008).

Parallel to the philosophical concept of rationality and the discussion on its relation with goals, there is literature both on rational goal achieving and on rational goal setting within the social sciences. Policy analysis sometimes distinguishes between a "rational model" of decision-making, based on rational choice theory and stressing the objective of social gain, and other models which stress that policy is both defined and implemented as an outcome of complex interaction between different interests and institutions; for example, incremental approaches stress the importance of continuity with previous policies, and group approaches stress the importance of compromise between conflicting interests (Dye 2008). While rationalism in policy analysis may have its merits under ideal circumstances, it ignores the role of individuals, the human factor, the possibility of erroneous predictions, and the costs of rational-comprehensive planning. Both for the purpose of understanding and evaluating policies, therefore, rational choice theory also needs to take the realities of the policy-making process into account. While social gain and justice are undoubtedly positive objectives, a realistic account of policy decision-making needs to take into account how such basic objectives are manifested in goals that reflect the compromises and restrictions of the political and sociotechnical environment. Such goals are instrumental in coordinating actions towards achieving the more general objectives of social gain and justice.

Some of the cruder arguments surrounding the philosophical dichotomy between instrumental and noninstrumental rationality can be avoided by observing the difference between intrinsic goals, which are pursued as ends in themselves, and goals that are set in order to coordinate actions over time or across agents. The former are typically more difficult to express and measure in a concrete or tangible way, like Aristotelian eudaimonia, and are arguably more difficult to assess as rational or irrational. Goals that coordinate action, or operationalize ends such as personal happiness or social justice, may more easily be subject to such assessment. In other words one may very well ask whether the goal to pursue a career in philosophy is rational in the sense that it would further a particular person's happiness or whether the objective to offset transportation costs associated with long distances for sparsely populated regions is rational in the sense that it would advance social justice, but the goals of happiness and justice themselves are not as easy to question directly. It is of course open to question what exactly constitutes happiness or justice and what exact actions to take in order to achieve these goals, and one may even ask whether happiness is the highest human good or whether justice is the first virtue of social institutions, but it would require sophisticated argumentation to determine whether happiness and justice are rational goals. The instrumental account of rationality therefore only makes sense for goals that are basic, in the sense that they are not set in order to achieve other goals. On the other hand, it makes sense for an agent to have nonbasic goals since such goals typically make it easier for the agent to coordinate actions towards achieving basic goals.

Continuing this argument, a goal that reflects the relevant compromises and restrictions of its context has the advantage that it can coordinate actions but the drawback that it is specific to this particular context. If the context changes, the goal may become less appropriate in the sense that it no longer captures the relevant compromises and restrictions that once made the goal a priority. If so, the actions coordinated by the goal become less instrumental in achieving the original intentions of the goal. This problem is especially notable when long-term perspectives are involved, as the context may change substantially over time. A goal that is open to several possible interpretations is more flexible in this sense. In the light of current demands for policies for sustainable development, it has been stressed that long-term policy goal formulation requires a balance between accounting for uncertainties by means of interpretative flexibility and learning processes on the one hand and the need to coordinate action on the other hand (Pinch and Bijker 1987; Voβ et al. 2009). Recent case studies of the problems involved in such long-term policy goal setting include transition management in the Netherlands, which aims to empower change agents, niches, and frontrunners, to reach a sustainable society (Avelino 2009; Kern and Howlett 2009).

While the philosophical literature is abundant within the wide topics of uncertainty and vagueness, as well as within the topic of responsibilities involving future possible worlds, philosophers have not given as much attention to combine their efforts from these topics into a normative theory of long-term goal setting. On such an account, lack of precision would be positive in the sense that future agents may interpret a goal better than present agents can do.

While increasing environmental concerns since the turn of the millennium have served to increase interest for goal flexibility and long-term adaptability, and while interventionist planning has learned from the lessons provided by the rise of (neo-) liberal ideology and the collapse of planned economies during the 1980s and the 1990s, theoretical studies on the process of policy goal setting go back at least to the early 1970s. Examples of policy goal-setting processes in earlier frameworks and models include feedback from policy outcomes as well as political debates and bargains (Amara 1972; Said 1974). From a policy research perspective, two inputs to the goal-setting process can be noted as relevant: the distribution of objectives and values in society and the societal impact of policy actions. Both are subject to change during time and to be fed back into the process during the course of policy implementation. Two crucial problems of relevance for the concept of rationality are the epistemological question of whether the distribution of values in society can be objectively known and the ontological question of whether social action is caused by social structures or by individual intentions (Dixon and Dogan 2004).

To summarize the discussion of this section and returning to the observations of the previous section, there is literature highlighting the dynamic nature of goals from both a descriptive and a normative perspective. Both goal setting and goal achieving are important aspects of rationality, and goals are formulated in the context and environment where they are achieved. This context may change over time, however, and it is important to formulate goals such that they can withstand such changes.

3 Normative Implications and Lessons Learned

In the following sections, the highlighted observations of the specific case are developed into general arguments. While the case of the Swedish transport policy goals is specific, it has already been noted that these goals have much in common with other national and transnational transport policy goals as well as some goals from other policy areas. These goals also illustrate topics of general philosophical interest.

3.1 Goals Are Subject to Evaluation and Updating

Since the 1960s, Sweden's national policy objectives have been updated about once every 10 years. The last revision was in 2009. The need to revise the goal system, from the previous system consisting in six separate sub-objectives to the current structure with a functional objective and an impact objective, was motivated by a number of reasons.[5]

Some of the Swedish transport policy objectives were overlapping – both conceptually and in terms of how they were to be achieved. The accessibility objective, for example, was difficult to completely distinguish from the quality objective.

[5] A comprehensive review of the recognized problems associated with the last version of the Swedish transport policy goals, and suggested improvements to mitigate these problems, was provided in reports 2008:2 and 2008:3 of the Swedish Institute for Communication Analysis. Some of these issues were recognized earlier and elaborated, for example, in reports 2000:1, 2000:5, and 2002:20 of the same organization.

Another example is provided by the quality and the safety goal; a transport system with a low risk of accidents or sabotage is also a reliable system with a low rate of disturbances. While such synergies are generally viewed as positive, as opposed to antagonisms or goal conflicts, they provide difficulties in the process of evaluating performance with respect to each individual objective.

The general topic of updating goals relates to the two theoretical distinctions mentioned in the previous section. Updating goals is a matter of goal setting and goal achieving, as well as of instrumental and noninstrumental rationality. The specific case of Swedish transport policy goals highlights a number of general lessons.

Goals that fail to coordinate actions, and goals that cannot be appropriately evaluated, will sooner or later need to be revised or abandoned. Such shortages can be conceptual, in which case they can be approached by adequately identifying the relevant distinctions and definitions required to clarify the goals, but difficulties in operationalizing goals may also depend on the context of the goals. While a goal may make perfect sense conceptually, the agent may face sets of alternatives and environmental constraints that make the goal inadequate for various reasons. Goal setting therefore typically needs to be preceded by a theory, which could be more or less elaborated, of how the goal can be achieved. Once the goal has been set, the agent will hopefully learn more about how the goal can be achieved, revise initial expectations or perceptions, and consequently develop the initial theory of how the goal can be achieved.

While goals provide the tools to evaluate actions, such evaluations also provide feedback to the initial setting of the goals. Goals may prove to be unachievable, irrelevant, or immeasurable – properties that are not desirable from an action coordinating point of view. Evaluations therefore serve not only to reveal progress and performance with respect to a given goal but also to reveal the achievability, desirability, and other important properties of the goal.

3.2 There Is a Trade-Off Between Precision and Flexibility

In Swedish transport policy, a recognized problem is that the overall objective of economic efficiency does not include everything usually associated with the idea of positive social development. This is also recognized in other policy areas as well as in theoretical discussions. Distribution of goods, for example, is irrelevant as far as Pareto efficiency is concerned, provided only that no person in society can become better off without someone else becoming worse off. Furthermore, sustainable development is not fully defined. The report of the World Commission on Environment and Development (Brundtland Commission 1987, p. 43) defines sustainable development as "development that meets the needs of the present without compromising the ability of future generations to meet their own needs", but the term "needs" will have different meanings for different people and it is unclear how strictly the term "compromising" ought to be interpreted. Furthermore, the bill proposing the Swedish transport policy objectives, referring to the UN conference in Rio de Janeiro in 1992, points out that sustainability should be "social, cultural, economic, and environmental". This generates four dimensions by which to determine "needs" and interpret "compromising". In the discussion anticipating the revision of the Swedish transport policy goals, it was presumed that ethical issues, distribution of goods, general matters of justice and fairness, and other general topics of relevance for transport policy but not directly covered by the concept of economic efficiency were covered by the sustainability concept. Given the multidimensional and polysemic nature of the sustainability notion, the conjunction of efficiency and sustainability has provided an overall goal with some flexibility.

As noted earlier, flexibility is positive for long-term goal setting, provided that this can be appropriately balanced with the need to coordinate efforts and actions. The adequacy of any goal changes with time. Unforeseen shifts in technology, opinion, and other relevant contextual attributes will be more likely the longer the time perspective. To manage such shifts, goals can be formulated in less precise terms such that the interpretation of the goal can shift correspondingly without losing continuity with the long-term strategy. Furthermore, the process of revising and updating goals will be slower the more complex the policy area and organizational structure are. Time and complexity are therefore important variables in identifying the balance between precision and flexibility.

Long-term policy is characterized by genuine uncertainty. While it is often difficult to identify likely contextual shifts in the long run, it may be possible to identify particular context areas of relevance where change is likely. It may be difficult to say exactly how the global economy or technological innovation will develop over the next 40 years, for example, but it is easier to say that these are relevant contextual areas where changes would have impact on how to prioritize environmental policy efforts. A reasonable strategy, therefore, is to formulate goals such that they provide flexibility in areas recognized as important for the policy in question but where development is uncertain.

While an imprecise goal can have the advantage that it provides flexibility, a certain level of precision is still required in order to be instructive and provide focus for action. Precision here means both the absence of multiple interpretations and the presence of clear performance indicators. Flexible goals therefore provide less action guidance than precise goals. On the other hand, precise goals require a more detailed theory of how they can be achieved. Generally, a precise goal is more context-dependent than an imprecise goal. An unforeseen change in the context can render a precise goal inappropriate. A key challenge to goal setting is therefore to balance the properties of precision and flexibility.

As an example, happiness maximization is a flexible but rather empty goal. Even though such a goal is distinct from other social goals, such as justice or equality, it is a goal that provides less action guidance for concrete policy decision-making than a more precise goal such as minimizing travel time for individual motorists. On the other hand, the goal of minimizing travel time may become less relevant with time as new forms of public transportation or communication technology emerges or as negative consequences of transportation such as environmental impacts receive more attention.

3.3 Different Kinds of Goals Require Different Approaches

Some of the first comprehensive reviews of the work to achieve the Swedish transport policy objectives from 1998 pointed out that the six sub-objectives were qualitatively different from each other. Many of these points still apply to the revised goal structure from 2009. While some of the objectives were mainly concerned with economic efficiency, other objectives are ethical or distributional. The objective of transport quality is the clearest example of the former. On the other hand, the traffic safety goal has a clear ethical dimension. Furthermore, the objectives of equal opportunities and regional development are mainly concerned with the distribution of benefits arising from the transport system. The objective of an accessible transport system is mainly seen as concerned with economic efficiency but also has a distributional element as it includes the condition that disabled persons should have access to the transport system. Noneconomic features – i.e. features that are not directly concerned with economic efficiency – of the policy objectives are mainly seen as aspects of sustainability. These qualitative differences between the Swedish transport policy objectives affect the adequacy of trade-off procedures; cost-benefit analysis gives more importance to economic objectives than to noneconomic objectives, and ethical objectives such as transport safety may be given a disproportionately low weight when only individual willingness to pay is taken into account (Tudela et al. 2006). It has on the other hand been pointed out that the safety and environmental objectives might push the other objectives aside because these two are more easily quantifiable than the others. In addition, the road safety goal is highly ambitious. Several writers have maintained that this goal seemingly implies an extreme priority for road safety (Elvik 1999), although the road safety goal itself makes no such explicit claim (Rosencrantz et al. 2007).

The function of the transport system has received relatively little attention. While the accessibility objective concerns the actual purpose of the transport system, accessibility is by itself a complex concept; as far as transportation is to function as a means of communication between individuals, accessibility requirements may be partially or fully offset by alternative solutions in telecommunication. The availability of online education or conferences, for example, affects the demand of travelling to attend meetings. Since transport is only valuable in the presence of travel demand, this poses problems for how to evaluate transport accessibility.

In line with the distinction between the functional objective and the impact objective of Swedish transport policy, it is interesting to stipulate a general distinction between functional goals and restraining goals. A functional goal is linked to a particular type of activities and the purpose of these activities. A functional objective is an objective to achieve something that is in the area of responsibilities or functions of a particular organization, policy, ability, or other entity with a defined purpose or use. Restraining goals, on the other hand, are linked to the activities of achieving functional goals and the undesired side effects of such activities. A restraining objective is an objective to minimize such undesirable side effects.

Minimization of undesirable side effects, as defined above, can be achieved in different ways. A proposed action to advance a functional goal may be turned down because of its negative side effects. A proposal for building a new road between two towns, for example, may be rejected as it is predicted to cause an increase in road traffic fatalities. Alternatively, the action may be taken but with additional actions taken in order to mitigate the side effects of the action. The road may be built, but with continuous improvements in road safety following its construction. Finally, alternative actions may be taken in order to reach the same functional goal without causing the negative side effects. Investments in other communication technologies may reduce the demand for transportation between the two towns in the example. The last possibility opens up for innovative solutions, but requires functionality comparisons of solutions that may be very different.

Another point relating to the distinction between functional and restraining objectives is that these two kinds of objectives are often justified in different ways. A functional goal may be justified by considerations of efficiency and progress, whereas a restraining goal may be justified by ethical considerations manifested in "responsible" or "acceptable" approaches to such progress. This suggests that different kinds of goals are associated with different kinds of attitudes among those responsible for implementing the goals as well as different kinds of rhetorical expressions in arguments advocating the goals.

4 Philosophical Relevance

The remaining part of this chapter discusses the relevance of the general arguments presented above for a number of selected topics from modern philosophical literature. The purpose is to explore implications of the claims presented above, in particular implications of relevance for the philosophy subject.

4.1 Future Generations

Goals imply obligations in the sense that they require agents to take actions to achieve them. Furthermore, as noted earlier, the setting of a goal is an act that depends on the agent's expectations of how the goal can be achieved. Ultimately, as evaluation of implementation efforts provides feedback to the agent, such expectations are formed by the agent's previous actions to achieve goals. There is an analogy between obligations to future generations and obligations to future goals; agents are obliged to respect the needs (or wishes, preferences, etc.) of persons whose existence will depend on actions today, and agents are obliged to achieve future goals that are the outcomes of implementation and evaluation processes today. In the literature on philosophical issues related to future generations, a fundamental paradox is that, while present persons are morally required to respect the needs of future persons, future persons owe their existence to present persons. Correspondingly, future goals will be formulated drawing on the lessons learned from the achievements and failures related to present goals. While the parallel between existence of future persons and formulation of future goals indicates that tools and results may be shared across these two topics, the parallel should not be taken too far. While both future generations and future goals generate future obligations and while both can be seen as outcomes of present actions, these two concerns are arguably different when it comes to present obligations. While the notion of sustainability requires that today's people respect the needs of future people, it is arguably not required from today's people that we should achieve future goals (if not interpreted as manifestations of future people's needs).

As for the trade-off between precision and flexibility, part of the rationale for providing flexible goals is that the relevant needs may change with time. Goals induce and coordinate actions, and goals should be realistic, constructive, and desirable throughout the implementation period. Hence, the balance between precision and flexibility should reflect uncertainties of what the needs of future people may be. However, the implementation period of a goal is typically relatively short. As mentioned earlier, transport policy goals are revised about every 10 years. Objectives for environmental protection are expected to guide policies for 40 years forward. In either case, the trade-off between flexibility and precision is more of an issue for the present generation or the next generation than for distant future generations. That is not to say, of course, that future generations are irrelevant to present goal setting, as such generations are affected by the impacts of present actions to achieve present goals. While present day goal setting needs to reflect priorities with respect to both present and future needs, the balance between precision and flexibility should reflect uncertainties in both kinds of needs.

Finally, concern for future generations can easily be perceived as a restraining goal. The notion that the needs of the present generation should be met without compromising the needs of future generations suggests that a negative side effect of meeting the needs of the present generation is that the needs of future generations are compromised. In other words, there is a functional goal to meet the needs of the present generation and a restraining goal to prevent that needs of future generations are compromised. Interpreted this way, however, the restraining goal is defined and measured in terms of the functional goal. That is, the restraining goal is actually a requirement that there is "sustainable functionality" or that the same level of functional standard should be possible to uphold in the long run.

4.2 *Standard of Measurement (Axiological Commensurability)*

In order to coordinate actions, ideally, performance should be possible to measure with respect to any goal. For any agent, furthermore, comparisons between performances with respect to different goals should ideally be possible to make. This means that the agent should be able to make at least some judgments regarding how much a certain level of performance with respect to one goal is worth in terms of performance with respect to other goals. Hence, establishing a standard of measurement is a key topic for performance evaluation. If measurement is perceived to be a problem and if failure to provide adequate performance measures results in failure to coordinate actions, then the goals should be revised; this does not necessarily mean that goals should be measured on the same scale, but prerequisites for performance evaluation should be made clear, and it should be clarified whether evaluative judgments may be suspended in cases where such a common standard of measurement is missing.

Vagueness is a way to express uncertainty about the standard of measurement in performance evaluation. There may be lack of consensus on how to establish such a standard, or on what the priorities should be. There may also be uncertainty regarding the impact of achieving different goals, for example, uncertainty on how much each goal contributes to the overall goal of a particular policy area. The longer the time period a goal is set to guide and coordinate actions, the more important it is to deal with such lack of consensus and certainty at the time of goal setting. The balance between precision and flexibility should reflect the level of consensus and certainty of the goal-setters.

As stated before, different kinds of goals (functional and restraining) are often justified in different ways. Different reasoning can result in failure to adequately provide a common standard of measurement, as different kinds of values are involved. For example, discussions on how to trade safety goals against other goals occasionally result in rhetorical questions on the value of human life. While some such lack of common measurement may be fundamental or unsolvable, other difficulties in providing a common standard of measurement may be overcome by analysing the justifications for each goal and making the reasons for setting these goals transparent.

4.3 Fairness

Equity is often associated with the notion of socioeconomic sustainability. The Swedish transport policy objectives provide an example of this; as distributive concerns are irrelevant as far as economic efficiency is concerned, but as this still is a relevant area of concern for Swedish transport policy, the combined goal of efficiency and sustainability suggests that distribution of benefits in society is captured by the notion of sustainability. According to this interpretation, the goals of a policy area will not be able to coordinate actions in the long run if they do not adequately address the issue of fairness. An unfair development is not a sustainable development, it is hypothesized, and goals that do not address such issues will not be accepted in the long run.

Interpretative flexibility of policy objectives provides space for shifting concerns and priorities for distributive justice. Fairness can take specific forms such as concern for future generations, disadvantaged regions, or gender equality, as well as non-equity-based aspects of fairness such as merit. These aspects of fairness have different significance for the policy in question and are affected in different ways by policy decisions. As policy goals are the outcome of compromise and are expected to be implemented during several mandate periods, lack of precision provides the flexibility to make the goals operationalizable until they are achieved.

Finally, fairness can be perceived as a restraining goal in general. For a specific policy area, this interpretation means that the functional goals are to be achieved in a fair way, that benefits of achieving the functional goals should be distributed fairly or that opportunities to influence actions carried out in order to achieve functional goals should be provided in a fair way.[6] While fairness in this way can be expressed in terms of functional goals, fairness remains a restraining goal to be balanced against the same functional goals.

5 Concluding Remarks

This chapter has used the case of Swedish transport policy objectives to illustrate three general points: First, goals are subject to evaluation and updating. Second, there is a trade-off between precision and flexibility. Third, different kinds of goals require different approaches. The implications of these three points for the philosophy subject have been explored within three separate topics: future generations, standard of measurement, and fairness.

References

Amara, R.C. 1972. Toward a framework for national goals and policy research. *Policy Sciences* 3: 59–69.

Avelino, F. 2009. Empowerment and the challenge of applying transition management to ongoing projects. *Policy Sciences* 42: 369–390.

Brundtland Commission (World Commission on Environment and Development). 1987. *Our common future*. Oxford: Oxford University Press.

Cingranelli, D.L., R.I. Hofferbert, and E.A. Ziegenhagen. 1980. Goal evolution through implementation: The problem for policy evaluation. *Policy Studies Journal* 8(7): 1229–1238.

Dixon, J., and R. Dogan. 2004. The conduct of policy analysis: Philosophical points of reference. *Review of Policy Research* 21(4): 559–579.

Dye, T.R. 2008. *Understanding public policy*, 12th ed. Upper Saddle River: Pearson Prentice Hall.

Edvardsson, K., and S.O. Hansson. 2003. When is a goal rational? *Social Choice and Welfare* 24(2): 343–361.

Elvik, R. 1999. Can injury prevention efforts go too far? Reflections on some possible implications of Vision Zero for road accident fatalities. *Accident Analysis and Prevention* 31(3): 265–286.

[6] In Swedish transport policy, gender equality and regional development is part of the functional goal.

Johansson, R. 2009. Vision Zero – Implementing a policy for traffic safety. *Safety Science* 47: 826–831.

Kern, F., and M. Howlett. 2009. Implementing transition management as policy reforms: A case study of the Dutch energy sector. *Policy Sciences* 42: 391–408.

Kuruvilla, S., and P. Dorstewitz. 2010. There is no 'point' in decision-making: A model of transactive rationality for public policy and administration. *Policy Sciences*. 43(3): 263–287. Available at www.springerlink.com/content/gp23hw6754p10511

Larsson, P., S.W.A. Dekker, and C. Tingvall. 2010. The need for a systems theory approach to road safety. *Safety Science* 48: 1167–1174.

Nagel, S.S. 1984. A bibliography on goals, means, and methods in public policy analysis. *Policy Studies Journal* 12(4): 777–778.

Parsons, W. 1995. *Public policy: An introduction to the theory and practice of policy analysis*. Cheltenham: Edward Elgar.

Pinch, T.J., and W.E. Bijker. 1987. The social construction of facts and artifacts: Or how the sociology of science and the sociology of technology might benefit each other. In *The social construction of technological systems*, ed. W.E. Bijker, T.P. Hughes, and T.J. Pinch, 17–50. Cambridge, MA: MIT Press.

Rawson, G.E. 1980. Organizational goals and their impact on the policy implementation process. *Policy Studies Journal* 8(7): 1109–1118.

Rosencrantz, H. 2008. Properties of goal systems: Consistency, conflict, and coherence. *Studia Logica* 82: 1–22.

Rosencrantz, H. 2009. Goal-setting and the logic of transport policy decisions. Royal Institute of Technology, PhD thesis, Stockholm.

Rosencrantz, H., K. Edvardsson, and S.O. Hansson. 2007. Vision Zero – Is it irrational? *Transportation research part A: Policy and practice* 41: 559–567.

Sabatier, P.A. (ed.). 1999. *Theories of the policy process*. Boulder: Westview Press.

Said, K.E. 1974. A policy-selection/goal-formulation model for public systems. *Policy Sciences* 5: 89–100.

Tudela, A., N. Akiki, and R. Cisternas. 2006. Comparing the output cost benefit and multi-criteria analysis – An application to urban transport investments. *Transportation Research Part A* 40: 414–423.

Voβ, J.-P., A. Smith, and J. Grin. 2009. Designing long-term policy: Rethinking transition management. *Policy Sciences* 42: 275–302.

Chapter 6
Rational Goals in Engineering Design: The Venice Dams

Karin Edvardsson Björnberg

1 Introduction

The rationality of scientific goals has been a much discussed topic in philosophy of science since the publication of Larry Laudan's Science and Values in 1984 (e.g. Iranzo 1995; Baumslag 1998; Címtora 1999). Until now, significantly less attention has been paid to the rationality of engineering goals, although exceptions exist (e.g. Hughes 2009; Kroes et al. 2009; de Vries 2009). As goals have a central action-directing and coordinating function in the engineering design process, there seems to be a gap in the research. Engineering projects usually start with an identified customer need or desire that is transformed into a set of functional requirements and design specifications for the development of the artefact. These needs, requirements and specifications serve as criteria for the development, testing, evaluation and readjustment of different design solutions. Negotiating and trading off different and often competing requirements is therefore an essential part of the engineering design process.

As goals (understood as customer needs, functional requirements and design specifications) have a central guiding role in the engineering design process, there is a need to investigate what normative criteria the goals should meet to be successful: What is a good (i.e. functional or rational) engineering goal?[1] In this chapter, a set of criteria developed by Edvardsson and Hansson (2005) and Edvardsson Björnberg (2009) as part of an evaluative framework for public policy goals is analysed using

[1] Throughout this chapter, the term "rationality" is given a wide interpretation. The term "rational engineering goal" is used to denote a goal that fulfils the typical function of goals to direct action in order to facilitate goal achievement. Readers who prefer a more restricted usage of the term "rationality" may instead use the terms "functional engineering goals" or "successful engineering goals" when references are made to "rational engineering goals."

K.E. Björnberg (✉)
Royal Institute of Technology, Stockholm, Sweden
e-mail: karine@kth.se

M.J. de Vries et al. (eds.), *Norms in Technology*, Philosophy of Engineering and Technology 9, DOI 10.1007/978-94-007-5243-6_6,

the large-scale engineering project of the Venice Dams (henceforth, the MOSE system) as an empirical basis. The fundamental assumption of the analysis is that goals that satisfy the set criteria fulfil their typical function as goals; they have the capacity to direct and coordinate action in a way that facilitates goal achievement (i.e. "achievement-inducing" goals).

Four criteria are particularly useful as analytical tools in the context of engineering design: precision, evaluability, approachability and consistency (see also Hughes 2009, p. 393).[2] Precision means that it is clear what the designed artefact should be able to achieve in terms of outer (customer-environment) and inner (device-environment) functions. Evaluability means that it is possible to establish the extent to which customer needs, functional requirements and design specifications have been met. Approachability means that it is feasible to come reasonably close to satisfying customer needs, functional requirements and design specifications. Consistency, or coherence, means that customer needs, functional requirements and design specifications can be met simultaneously; hence, that there are no significant conflicts among the requirements. The normative strength of the goal criteria varies depending on which part of the engineering design process is under consideration. As a general rule, the "further down" in the transformation process from customer needs to design specifications, the more important it is that the goals engineers work with satisfy the identified criteria.

Prior to the philosophical analysis, the role of goals in engineering design is discussed. The function and design of the MOSE system are also described in greater detail. This is done in Sects. 2 and 3, respectively. In Sects. 4, 5, 6 and 7, the four goal criteria are analysed using examples from the MOSE design process. Section 8 contains some concluding remarks. Contrary to the idea, embraced by many philosophers, that goals unlike means cannot be assessed in terms of rationality, this chapter concludes that engineering goals can indeed be evaluated on the basis of such considerations.

2 The Function of Engineering Goals

Engineering projects start with an identified customer need or problem and end with the production of an artefact or artefact-based service that satisfies this need or solves the problem.[3] Between the problem-identification stage and the production stage, a variety of activities take place. Some of these activities are referred to as

---

[2] Edvardsson and Hansson (2005) indicate that a goal should ideally also satisfy the criterion of motivity, that is, it should have the capacity to motivate action that facilitates goal achievement. This criterion is not discussed at any length in this chapter because it has a subordinate role in engineering design. Engineers work professionally and are therefore committed by external forces to act in ways that further goal achievement; hence, the motivation to do so is not triggered by the goals themselves.

[3] The fact that engineering projects start with an established customer need does not necessarily mean that an actual customer has expressed this need. Sometimes, engineering design projects proceed from the engineers' own estimations of what customers (or the market) desire or can be made to desire (de Vries 2009, p. 494).

engineering design. The Accreditation Board for Engineering and Technology (ABET) defines engineering design as:

> [t]he process of devising a system, component, or process to meet desired needs. It is a decision-making process (often iterative), in which the basic sciences and mathematics, and the engineering sciences are applied to convert resources optimally to meet these stated needs.[4]

The outcome of the engineering design process is a "use plan", or description, of a technical artefact that contains everything that must be known in order to manufacture that artefact (Kroes 2009, p. 513). The use plan can be supplemented with a prototype that demonstrates the effectiveness of the plan. As there is wide variety in engineering design practice, there are many kinds of artefacts, design problems and designer competencies. However, common to all design practices is that they are decision-making activities (Kroes et al. 2009). The engineering design process essentially proceeds through a large number of decisions: how many design options to generate, which options to develop further and which to abandon, how to make trade-offs between conflicting design requirements, what criteria to use in the evaluation of design solutions, etc.

The starting point of most design projects is an identified customer need.[5] In the engineering design process, customer needs are translated, or transformed, into functional requirements (de Vries 2009). Functional requirements describe what the artefact must be able to do in order to satisfy the identified customer need. For example, for a flood barrier, the functional requirements could be capability of safeguarding from tides up to a certain level, capability of letting commercial and rescue vessels pass through the barrier while in operation and capability of being fully raised within a certain time from a received warning. Functional requirements are largely customer oriented and say little about the internal structure of the artefact, for example, about its geometrical, physical or chemical properties (Kroes 2009, p. 524).

Beside functional requirements, there are many other concerns to be addressed in the engineering design process: technical, environmental, economic, recreational, safety, aesthetic, etc. To take a simple example, a flood barrier should not only be able to safeguard the coast from flooding; the barrier should also be able to do so without causing damage to the environment and without jeopardising safety and recreational values. In the engineering design process, these concerns are often expressed through constraints on how the artefact should be designed and implemented. Some constraints are formal, that is, written out in regulations, engineering codes, contracts and other documents, whereas others are informal or only explicitly articulated when challenged (Bucciarelli 1994, p. 131). Some constraints are negotiable within limits, and others

[4] Criterion 5(b), ABET Criteria for Accrediting Engineering Programs 2009–2010, www.abet.org (accessed 10 April 2010).

[5] The customer can, for example, be an individual or a group of individuals, a public or private organisation, a company, the general public or the market. In large-scale engineering projects, such as flood barrier construction, the customer is often a public or semiprivate organisation that is formally authorised to represent a particular community or the general public.

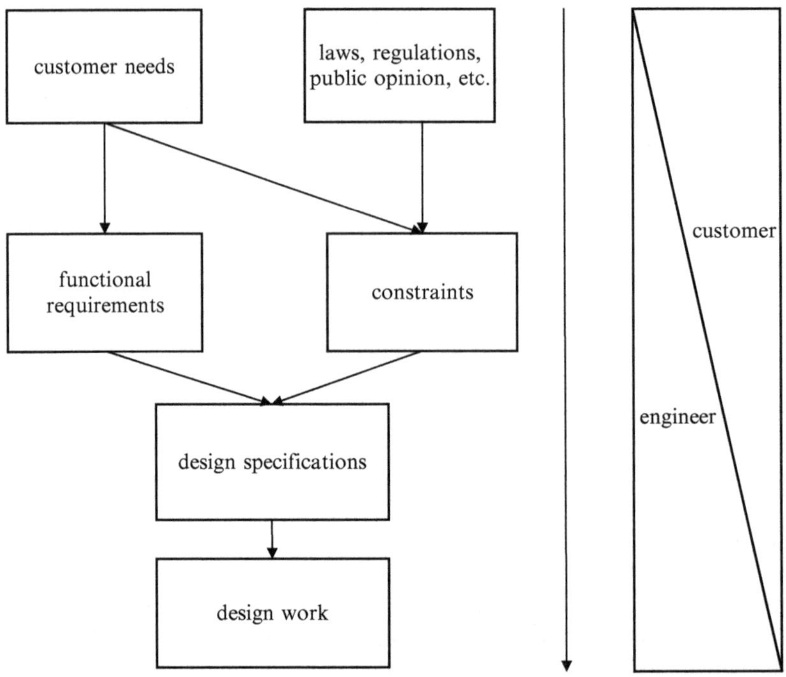

Fig. 6.1 The design process

are not. In large-scale engineering projects, constraints are often expressed both through formal documents, such as laws, regulations and environmental impact assessments, and informally, for example, through public opinion.

The distinction between functional requirements and constraints is not clear cut.[6] Customer needs reflect a multitude of values that not only concern *why* but *how* a particular artefact should be developed. The reason why flood barriers are built is that coastal areas need to be protected. However, coastal protection is not an unambiguous goal. It can be achieved in ways that differ in environmental, economic and social respects. Few would agree to build a barrier that effectively solves the problem of flooding but at the same time causes irreversible or catastrophic environmental damage. In the standard model of engineering design, customer needs are expressed through two separate categories of norms: functional requirements and constraints. However, in terms of reflecting what the customer wants, functional requirements and constraints cannot be separated in this idealised way (de Vries 2009, p. 497) (see Fig. 6.1).

[6] Consider mobile phones and laptops. These have been developed because people need to be able to make calls and do work even when they are not physically in their offices. Being portable is thus a functional requirement; hence, the weight of a mobile phone or a laptop is a functional requirement that can be expressed in physical terms, for example, "x should not weigh more than". I am grateful to Sven Ove Hansson for pointing this out.

Taking constraints into consideration, functional requirements are transformed into design specifications. Design specifications are structural descriptions of the artefact that specify measurable properties and behaviours. They are oriented towards the inner environment of the artefact – the artefact's internal structure – and say nothing about the function or why the artefact should be produced (Kroes 2009). The transformation process from functional requirements to design specifications is not a one-to-one projection; one functional requirement can be related to more than one physical feature, and vice versa (de Vries 2009, p. 498). If the set of design specifications adequately meets the functional requirements and the functional requirements are adequate operationalisations of the expressed customer needs, then the designed artefact is a reliable means to satisfy these needs.

Because engineering design problems are often "ill-defined", the design process is seldom linear.[7] Functional requirements, design specifications and constraints often need to be readjusted as the design process proceeds. Some requirements, specifications or constraints may turn out to be superfluous; others may need to be added as knowledge is gained about the design problem. Therefore, as noted by Kroes (2009), engineering design is not so much a linear problem-solving activity starting from a fixed set of goals as a decision-making process that proceeds iteratively and through feedback loops. In the decision-making process, setting and fine tuning of goals is equally as important as finding the appropriate means of achieving goals.

Goals have a central normative function in engineering design. Customer needs transformed into functional requirements serve as the broad goals of a design project. They are used as a normative standard for a variety of decisions, perhaps most notably for the generation, evaluation and prioritisation of design solutions (specifications), but also for more practical decisions, for example, which competencies to involve in the project and how to allocate resources between different engineering teams. Design specifications, in turn, are used to direct action in a way that fulfils the functional requirements of an artefact and serve as normative standards for actual construction work.

The goals that engineers work with usually direct action both intertemporally (over time) and interpersonally (between different engineers or engineering teams). Intertemporally, goals enable engineers and engineering teams to plan their activities over time so that the goals are more easily achieved. Goals introduce a certain settledness about action and deliberation. The adoption of a goal means that

---

[7] According to Cross (2000, pp. 14–15), ill-defined design problems are characterised by the following:

1. The design problem is vaguely defined (i.e. goals are ambiguous, many constraints and criteria are unknown, and the problem context is poorly understood).
2. Any problem formulation may contain inconsistencies.
3. Formulations of the design problem are solution dependent (i.e. it is difficult to formulate the design problem without referring to a solution concept).
4. Searching for design solutions is a means of understanding the design problem.
5. There is no definitive solution to the design problem.

there is commitment and that action will be taken to achieve it (McCann 1991, p. 206). This holds true in the engineering context too. Having decided on a particular goal, an engineer does not consider design solutions, plans and strategies that clearly work against the goal, at least until the goal itself has been reconsidered (cf. Levi 1986, p. 69ff.; Nozick 1993, p. 146). The goal functions as a "conduct controller" in that it (at least temporarily) narrows down the engineer's scope of future deliberations to a limited set of design solutions and provides a reason for considering some solutions but not others (cf. Bratman 1999; Schmidtz 1995, p. 7).

Engineering goals also facilitate coordination between different engineering teams. On the basis of an adopted goal, different engineers and engineering teams can beneficially plan and coordinate their work. The planning benefits gained in these situations also are derived from the stability of action and deliberation conveyed through goals. When an engineering team has adopted a particular goal, other engineering teams can, to some extent, predict what type of design solutions the first engineering team will search for, or at least what design solutions the team will not consider and adjust their own efforts accordingly (cf. Nozick 1993, p. 9).

3 Designing the MOSE System

The MOSE system (acronym for "Modulo Sperimentale Elettromeccanico", Experimental Electromechanical Module) is a set of mobile barriers designed to defend the City of Venice from high waters.[8] The barriers, comprised of flap gates built into the seabeds of the Venice lagoon, are placed at the Lido, Malamocco and Chioggia inlets. When a tide exceeding 110 cm is forecast, the barriers are emptied of water by emission of compressed air. This causes the barriers to rise until they emerge above sea level and isolate the lagoon from the sea (Fig. 6.2). Depending on the severity of the situation, the MOSE system can close one inlet at a time or all inlets simultaneously. The barriers remain risen until the high tide has passed (approximately 4–5 h) and are then "folded away" into their housings at the bottom of the inlets. A navigation lock at the Malamocco inlet allows commercial ships to pass the barrier even when the mobile gates are raised. Small navigation locks are also placed at the Lido and Chioggia inlets.

The history of the MOSE project dates back to the early 1970s. After the dramatic floods of November 1966, the water issue in Venice began to attract increased attention. It became obvious that anticipatory action was needed to safeguard Venice from flooding. The problem was formally recognised in 1973 in Law No. 171/73, which declared the water issue in Venice to be "of priority national interest". Following the adoption of the law, the Ministry of Public Works issued a call for tender for the design and construction of high water defences for the Venice lagoon.

[8] Information about the MOSE project and other interventions to safeguard Venice and the Venice lagoon can be found on the Consorzio Venezia Nuova website (www.salve.it).

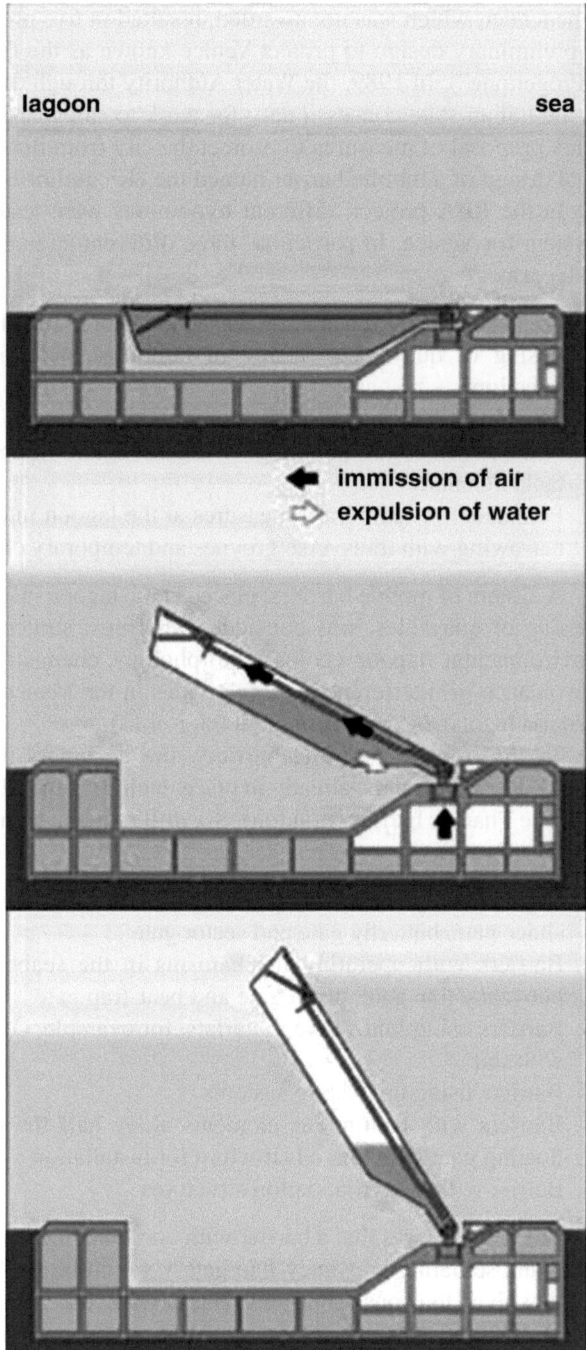

Fig. 6.2 The MOSE floodgates in operation (www.salve.it)

The tender, which was not awarded, resulted in five projects being used to draw up a preliminary design to protect Venice known as the General Feasibility Plan (the "Progettone"). In 1989, the Water Authority through the Consorzio Venezia Nuova – the Italian state concessionary for work to safeguard Venice – presented a complex proposal of measures to protect the city from flooding, including the conceptual design of a mobile barrier named the "Riequilibrio E Ambiente", REA.

In the REA project, different hypotheses were tested to find the best defence system for Venice. In particular, three different strategies were considered (www.salve.it):

1. Protection of built-up areas in the lagoon with local defence measures, for example, raising of quaysides, raising of buildings and raising of the land by deep injections
2. Modification of the physical structure of the lagoon, for example, reconstruction of the raised edges along lagoon channels and filling in of the Canale Petroli oil tanker channel
3. Permanent or temporary measures at the lagoon inlets, for example, permanent narrowing with transverse groynes and temporary closure of the lagoon inlets

A system of mobile barriers, placed at the lagoon inlets in combination with local raising of quaysides, was considered the most strategic solution considering the environmental (lagoon ecology, morphology, chemistry, etc.) and socio-economic (avoidance of interference with activities in the Venice ports, etc.) constraints and criteria involved (Spencer et al. 2005, p. 143).

To find the best mobile barriers, the designers then analysed a number of flood-defence systems already in place, including the Maesland barrier (Rotterdam) and the Thames barrier (London). Six different conceptual designs were considered (www.salve.it):

1. Barriers with intermediate piles to install the operating mechanisms, for example, sluice gate, butterfly gate and sector gate
2. Barriers with operating mechanisms in the seabed, for example, oscillating buoyancy flap gate, drum gate and bear-trap gate
3. Barriers using innovative materials, for example, cushion gate and floating gate with sail
4. Barriers using innovative systems
5. Barriers with continuous elements along half the inlet channel, for example, floating gate with hinged structure for installation
6. Barrier with deep foundation structures

The decision was that a barrier with operating mechanisms in the seabed – in this case an oscillating buoyancy flap gate – would be the best solution for the Venice lagoon given the constraints and criteria involved.

In 1992, after extensive experimentation, the preliminary design for the mobile barriers was completed. A negative response to the barrier design by the Ministry of the Environment's Environmental Assessment Commission in 1998 resulted in an update and further developments. In 2002, the final design of the MOSE system was presented by the Consorzio Venezia Nuova.[9] In April 2003, the formal decision to build the barrier was taken by the Committee for Planning, Coordination and Control ("Comitatone") in Rome.

Evident in the MOSE project is that engineering design is essentially an iterative decision-making process that often starts out from an "ill-structured" problem and operates under a wide variety of constraints. It became clear that the initial design remit was ill-structured as the Public Works Administration evaluated the proposals presented in response to the 1975 call for tender. As the evaluation process continued, it became obvious that none of the proposals could be selected. By the time the evaluation process came to an end, the design problem was revealed in its full complexity. Before a solution satisfying the manifold requirements, constraints and criteria involved could even be suggested, in-depth, multi-sectoral and multidisciplinary knowledge and information about the situation in the lagoon was needed (www.salve.it).

The MOSE design process operated under a wide variety of constraints; this is evidenced in the many documents issued by local and national authorities during the process. For example, in Vote no. 209 of 1982, the Higher Council of Public Works formulated a set of instructions and recommendations (later incorporated into Law 798/84) that had to be taken into consideration. The instructions emphasised the importance for any design project of "[s]afeguarding objectives and characteristics inherent in the unity and physical continuity of the lagoon" (www.salve.it). The instructions also highlighted the need to proceed through an iterative decision-making process, that is, gradually and experimentally, and in close collaboration with research.

4 Precision

Engineering design goals need to be precise, not only to guide the engineer(s) responsible for the implementation of the goals but also to constitute useful criteria on the basis of which proposed design solutions (conceptual designs, prototypes, etc.) can be assessed and evaluated. Only when engineering design goals are precise "[c]an one test the design against the requirements and determine whether or not the former satisfies the latter" (Maimon and Braha 1996, p. 146).

There are different types and degrees of goal specification. A goal is directionally precise if it specifies a direction for the engineer's work – this is a straightforward minimum requirement for goal specification. A goal has completive precision if it specifies not only in what direction the engineer should go to reach the goal but also to what extent the goal should be realised. For example, the goal "After closing the barriers, Venice should suffer much less risk of flooding" has directional precision, whereas the goal "After closing the barriers, the rate of rise of the lagoon water level

[9] Construction details of the barrier are described in Eprim (2005).

due to inflow through the gaps should not exceed 3 mm h-1" has directional and completive precision.[10] Finally, an engineering design goal is temporally precise if it specifies a point in time when the goal should be achieved.

Actual degrees of goal specification vary between the different levels of the design process. As a general rule, the "further down" in the transformation process from customer needs to design specifications, the more precise goals tend to be. Customer needs usually have only directional precision, at least in the initial stage. This is part of the reason why engineering design problems are often "ill-defined."[11] Sometimes, the customer knows what type of artefact is wanted but has little idea of possible designs. At other times, the need is much vaguer and is expressed simply as a problem that must be solved (Cross 2000, p. 61). This was, for example, the case in the MOSE project, where the starting point for the engineers' work was an identified problem to be solved but with little or no indication of what sort of solution would be approved in the end.

At the design-specification level, the goals that engineers work with are usually well defined. Often, the goals are formulated in quantitative terms that specify acceptable limits concerning size, weight, pressure, durability, temperature, noise, etc. In large-scale engineering projects, there are a great number of design specifications. Work is usually divided between different engineering teams, and separate sets of design specifications are made for different parts of the artefact. In the MOSE project, these included the foundation caissons, the gates, the electrical services, the preparatory works, etc.

For an engineer, goal imprecision can be a problem at all stages in the design process. Vaguely expressed customer needs are difficult to operationalise into functional requirements, and vague functional requirements are difficult to transform into adequate design specifications. In actual design practice, imprecision often creates difficulties in the problem-definition phase, that is, when customer needs are transformed into functional requirements. There is a necessity to obtain a satisfactory definition of the problem. However, because customers and engineers express themselves differently, problem definition is a difficult task. A word used by the customer to express a need or describe a particular problem may have a different meaning for the engineer, or may not be precise enough to guide the engineer's design work (de Vries 2009, p. 495). To arrive at a common definition, the transformation from customer needs to functional requirements often has to proceed iteratively and through negotiation.

Engineering goals that are broad or vague are generally problematic from an action-guiding standpoint since they leave the engineer with little idea of what to aim for. Vague customer needs or functional requirements can lead to inappropriate solutions being considered and then discarded when it is found that they do not fall within the range of what is considered acceptable (Cross 2000, p. 91). This can make the design process unnecessarily expensive. On the other hand, well-defined

[10] The last goal formulation is taken from Eprim (2005, p. 257).

[11] See footnote 7.

engineering goals can also be problematic. Detailed customer needs have the potential to hamper the creativity of the engineer and lead to designs that are inefficient from a multi-goal perspective. When the problem definition is narrow, the engineer may overlook unusual or unconventional solutions that could otherwise help to solve the problem (Dieter 1991, p. 7). It is therefore preferable to formulate the overall goals of an engineering project as broadly as possible at first and then narrow them down at a later date if necessary (Cross 2000, p. 78).

5 Evaluability

To guide action, engineering design goals must not only be precise. Goals should also be evaluable, that is, it should be possible to measure degrees of goal achievement. Only when goals (and criteria) are evaluable is it possible to assess the adequacy of different design proposals before deciding on a final version for manufacture (Cross 2000, p. 6). Similarly, final design solutions can only be properly assessed when the goals and criteria against which the solutions are evaluated are transparent. Furthermore, evaluability guarantees that accountability for insufficient goal achievement can be established, at least in principle. This is particularly important in large-scale engineering projects with a number of engineering teams.

Successful goal evaluation presupposes that the intended end state is adequately specified. This requirement is basically met when engineering goals are precise as previously described. However, since evaluation is an assessment of the outcome in light of adopted goals and criteria, successful goal evaluation also presupposes that it is possible to establish how well the engineer's design solution aligns with the intended end state (Edvardsson and Hansson 2005, p. 353). In other words, degrees of partial attainment must be distinguishable.

To successfully measure goal achievement, at least two things must be decided: (1) what indicators should be used to signal goal achievement and (2) what levels of attainment (for these indicators) should be used to signal sufficient goal achievement. If the goal is to design a barrier that protects Venice from floods while permitting a functional lagoon ecosystem to thrive, for example, different sets of environmental indicators could be used to signal goal achievement. These include levels of erosion/sedimentation in the lagoon, levels of toxic substances, occurrence of mercury cycling, presence of pathogenic microbial communities in water or sediment and occurrence of salt marsh colonisation (Deheyn and Shaffer 2007, p. 208). For each of these, it must then be decided what levels signal satisfactory goal achievement. These choices are not value neutral. Deciding on a particular set of indicators or level of attainment indicates which aspects of the goal are most important.

Evaluability is usually not a problem at the design-specification level when goals are quantitative and well defined. Given that there are reliable measurement instruments and methods available to test a design solution, assessing goal achievement is for the most part a straightforward activity. However, assessing goal achievement at "higher levels" in the engineering design process is challenging. Two factors complicate the evaluation of design solutions. First, goals at "higher levels" in the engineering design process are sometimes vague, abstract or value-laden. To save Venice from floods while permitting a functional lagoon ecosystem to thrive is an example of an abstract and value-laden goal since there is no absolute definition of what constitutes a "functional lagoon ecosystem". When abstract or value-laden goals are operationalised into quantitative design specifications, indicators and criteria, there is always a risk that the engineer could lose sight of the overarching goals. The various operationalisations of the overarching goals instead become the "real goals" of the engineer's project.

Second, design solutions generally have to be evaluated against multiple goals or criteria. Different design solutions advance different goals to different extents. For example, one particular type of flood-barrier design could better satisfy environmental considerations compared with other designs while at the same time be more troublesome from the viewpoint of safety and commercial activity. An important part of the evaluation process is weighing these different goals, or criteria, against one another. There is a question as to whether any rational method can be employed to find a design solution that is optimal from a multi-criteria perspective. Philosophers of technology have argued against the existence of such a rational method (e.g. Franssen and Bucciarelli 2004; Franssen 2005). Drawing on insights gained in social choice theory, they argue that since no general rational solution method can be found for multi-criteria decision problems of this kind, it is problematic to talk about designs as optimal solutions. Instead, engineering design should be understood as a social process in which different goals and interests are negotiated and in which bargaining and strategic thinking are recurrent phenomena (Franssen et al. 2009).

6 Approachability

That goals should be attainable, or at least possible to reach to a meaningful degree, is a criterion often referred to in discussions of goal-setting rationality (Edvardsson Björnberg 2008). For example, in relation to scientific goals, Laudan (1984) has argued that only those goals that satisfy a particular set of conditions – namely, those goals that are not demonstrably, semantically or epistemically utopian – should be accepted as valid goals. His argument is that unrealistic goals should be rejected because it is unreasonable to adopt goals that are of no use in the selection of means towards their realisation. The normative criterion of approachability has also attracted some attention in engineering design literature. In actual design practice, the criterion is explicitly taken into consideration via feasibility studies (Hughes 2009). Feasibility studies are conducted – usually on conceptual designs – in order to validate the (technical, economical, legal, etc.) feasibility of a proposed solution and to make sure that the design that is eventually chosen satisfies customer needs with the available resources (Ertas and Jones 1996, p. 11).

Engineering goals can be divided into goals that represent demands and goals that represent wishes or wants (Pahl and Breitz 1996, p. 131). Demands are minimum requirements that must be met. For obvious reasons, such goals need to be fully attainable. Wants are aspirational targets that the customer or designer would like to meet as far as possible and are therefore worth incorporating into the design project (Cross 2000, p. 94). Ideally these targets should be possible to reach to some degree, but do not have to be fully attainable.

The degree to which an engineering design goal is attainable is often determined by the relations that exist between the goal and other goals. Some design goals may be realistic in that they are possible to attain or approach but only at very high cost. Such goals are only realistic in a weak sense. In actual design practice, degrees of attainability often relate to the financial resources available. For example, in flood-barrier design, it might be possible to reconcile all effectiveness, safety and environmental considerations through some innovative solution, but this would most likely come at an extreme and indefensible cost.

How ambitious should the goals set by engineers be? In software engineering, it has been argued that unrealistic goals, such as designing data programmes entirely free of bugs, should not be adopted since overly strong responsibility for goal failure would then accrue to the engineers who design the programmes (Stieb 2008; see also Miller 2008; Huff 2008). From a moral viewpoint, accepting such responsibility goes beyond what can reasonably be expected from a software engineer.

From a rationality viewpoint, however, there is evidence to support the argument that engineering goals should be relatively challenging. Psychological research has found that there is a positive correlation between the degree of difficulty pertaining to a goal and the level of performance displayed by the agent as they try to reach the goal.[12] The relationship is sometimes referred to as the "goal-difficulty function".[13] In simple terms, the goal-difficulty function implies that the more challenging a goal, the greater effort the agent is likely to make in order to reach the goal. This is believed to hold true at least up to a certain level of goal difficulty. For goals that are perceived to be very ambitious or challenging, though, research has shown that the agent may lose motivation to reach the goal (Stedry and Kay 1966). The latter argument is that when an unrealistic goal is used as a referent to which the agent's performance is compared, the discrepancy between the goal and the agent's actual performance will be so great that the goal does not, as with realistic goals, create corrective motivation to change the agent's behaviour (Campion and Lord 1982).

Based on the former evidence, it could be argued that challenging rather than easily attained goals should be set since this would likely lead to higher levels of goal achievement, that is, better designs. Setting ambitious goals, or sets of goals that are difficult to co-achieve, could motivate engineers to present creative and unexpected solutions that would otherwise not have been discovered. In other words,

[12] However, to the author's knowledge, no such research has been undertaken on engineering goals.

[13] A comprehensive meta-study of the empirical research on the goal-difficulty function can be found in Locke and Latham (1990). See also Locke and Latham (2002) for a shorter but more recent discussion.

highly demanding engineering goals or ideals could be an effective tool for "approximating the good" (Huff 2008, p. 243).

However, because engineering design problems are often ill-defined, it can be difficult to set sufficiently challenging engineering goals. When there is no definitive problem formulation at the outset, complexities and inconsistencies can cause the initial problem to be poorly understood, as in the MOSE project. Therefore, engineering goals have to be set on a somewhat intuitive basis and then be subsequently adjusted as more knowledge is gained. This is particularly the case in large-scale engineering where an often complex combination of factors determines the success or failure of a particular design.

7 Consistency

To successfully guide action, engineering goals also need to be consistent. A set of goals is consistent when measures can (realistically) be taken to approach one goal without making it more difficult to achieve the other goals in the set. Conflicting goals guide action in opposite directions, making it more difficult to coordinate action effectively among engineers and engineering teams. In the absence of clear goal prioritisation and a continuous dialogue between the design teams involved, the design project can become costly and time-consuming.

However, though generally problematic from a planning perspective, it may not be irrational to adopt conflicting goals. There are several reasons why this holds true in an engineering design context. Firstly, it can be difficult to know in advance the extent to which a set of engineering goals or requirements are consistent. Consistent goals that are adopted at one point in the design process may conflict when more detailed information about feasibility becomes available or when conditions external to the goals change – constraints may be added, environmental conditions may change, the financial situation may deteriorate, etc. (Edvardsson Björnberg 2009). For example, the on-going global financial crisis could make some of the more attractive and expensive design solutions in the Venice Dams case impossible to implement in practice. Conversely, goals that were previously in conflict can become more consistent as finances improve or new design solutions (materials, processes, etc.) are discovered.

Second, to create sets of goals that are free of conflict, engineers would have to systematically avoid goals that represent any real challenge or commitment since there is always a risk of conflict.[14] This is not desirable, as noted, since working towards goals that are difficult to reconcile can spur innovative reflection, discussion

[14] An analogous argument has been made by Hansson (1998), who argues against the position that a person's moral values always need to be consistent. The reason for this is that strategies that reduce the incidence of moral dilemmas tend to have side effects that are not worth the price. For example, to avoid dilemmas as much as possible, a person would have to systematically avoid commitments to other people since virtually any such commitment increases the risk that they will later be caught in a dilemma.

and action (see Sect. 6). An example is the attempt to reconcile the goal of effectively protecting Venice against flooding with stated goals such as preserving the "unity and physical continuity of the lagoon". Although difficult, such challenges have allowed design practice to evolve successfully.

Third, in engineering practice, goal conflicts are more or less considered a given. In large-scale engineering design, safety and economic requirements could conflict, as could economic with environmental or ethical requirements. There is no easy way to avoid this; negotiating requirements and finding solutions that reconcile conflicting goals are considered integral to engineering design.

Different methods can be used to solve or negotiate engineering goal conflicts; however, these are not discussed at any length in this chapter.[15] There are basically two levels at which such goal conflicts can be dealt with. Goal conflicts can be approached in the problem definition phase, as customer needs are transformed into functional requirements. In practice, this means that adjustments and compromises are made before the goals of the design project are set (cf. Rosencrantz et al. 2007). The other approach is for goals to be set and then negotiated at "lower levels" in the design process. As a general rule, the "further down" in the design process the negotiation takes place, the greater influence engineers usually have over the goals and value trade-offs involved (see Fig. 6.1).

8 Concluding Remarks

This chapter argues that engineering goals can be meaningfully evaluated in terms of rationality. Four criteria – precision, evaluability, approachability and consistency – have been analysed in relation to the MOSE system engineering project in Venice. The argument has been that although all criteria are relevant in a goal-setting context, the normative strength of the criteria varies depending on which part of the engineering process is under consideration. As a general rule, the "further down" one is in the transformation process from customer needs to design specifications, the more important it is that the goals with which engineers work satisfy the identified goal criteria.

It should be noted that the analysis in this chapter has limited scope and that further research is needed to establish the conditions for effective goal setting in engineering design. In this chapter, only four goal criteria have been analysed. In addition, there may be a need to identify other criteria in order for an engineering goal, or a set of engineering goals, to be rational. Examples of such criteria could be that engineering goals are understandable and communicable (which is not equivalent to being precise) or that sets of engineering goals consist of a limited number of goals.

Furthermore, for an engineering goal to be considered rational, it may not be sufficient to analyse the structural properties as has been done in this chapter. There

[15] For a description of some of these methods, see Chapter 10 in Cross (2000) and de Vries (2009).

should also be an analysis of the goal-setting process, for example, to determine if goals were set in accordance with established design methodologies or if the arguments that were used in this process were based on sound scientific or technological knowledge.

Finally, there is the argument that to function well, it is not sufficient that engineering goals only meet the goal criteria presented in this chapter; the goals must also operate in a "supportive" organisational context. This could, for example, mean that there is a well-defined system of monitoring, evaluation and feedback in force at all stages in the design process to ensure that knowledge and experience can be shared among the project engineers. This, in turn, presupposes that there are adequate channels of communication among different engineering teams and between the design project's leadership and the engineers responsible for day-to-day functions.

Acknowledgments I would like to thank Professor Marc de Vries, Professor Sven Ove Hansson and Professor Peter Kroes for their valuable comments and suggestions. I would also like to thank the participants at Track 7: Philosophy of Engineering and Design at the 2011 Society for Philosophy and Technology (SPT) conference in Denton, Texas, for their comments. Any remaining errors, if any, are mine.

References

Baumslag, D. 1998. Choosing scientific goals: The need for a normative approach. *Studies in History and Philosophy of Science* 29(1): 81–96.

Bratman, M.E. 1999. *Intention, plans, and practical reason*. Stanford: CSLI Publications.

Bucciarelli, L.L. 1994. *Designing engineers*. Cambridge, MA/London: The MIT Press.

Campion, M.A., and R.G. Lord. 1982. A control systems conceptualization of the goal-setting and changing process. *Organizational Behavior and Human Performance* 30(2): 265–287.

Címtora, A. 1999. Critical comments on Laudan's theory of scientific aims. *Sorites* 10: 19–38.

Cross, N. 2000. *Engineering design methods: Strategies for product design*, 3rd ed. Chichester: Wiley.

de Vries, M.J. 2009. Translating customer requirements into technical specifications. In *Handbook of the philosophy of science. Vol. 9: Philosophy of technology and engineering sciences*, ed. A. Meijers, 489–512. Amsterdam: Elsevier.

Deheyn, D.D., and L.R. Shaffer. 2007. Saving Venice: Engineering and ecology in the Venice lagoon. *Technology in Society* 29: 205–213.

Dieter, G.E. 1991. *Engineering Design: A Materials and Processing Approach*, 2nd ed. New York: McGraw-Hill.

Edvardsson Björnberg, K. 2008. Utopian goals: Four objections and a cautious defence. *Philosophy in the Contemporary World* 15(1): 139–154.

Edvardsson Björnberg, K. 2009. What relations can hold among goals, and why does it matter? *Crítica Revista Hispanoamericana de Filosofía* 41(121): 47–66.

Edvardsson, K., and S.O. Hansson. 2005. When is a goal rational? *Social Choice and Welfare* 24(2): 343–361.

Eprim, Y. 2005. Venice mobile barriers project: Barrier caissons construction details. In *Flooding and environmental challenges for Venice and its Lagoon: State of knowledge*, ed. C.A. Fletcher and T. Spencer, 257–262. Cambridge: Cambridge University Press.

Ertas, A., and J.C. Jones. 1996. *The engineering design process*, 2nd ed. Chichester: Wiley.

Franssen, M. 2005. Arrow's theorem, multi-criteria decision problems and multi-attribute preferences in engineering design. *Research in Engineering Design* 16(1–2): 42–56.

Franssen, M., and L.L. Bucciarelli. 2004. On rationality in engineering design. *Journal of Mechanical Design* 126(6): 945–949.

Franssen, M., Lokhorst, G.-J., and I. van de Poel. 2009. Philosophy of technology. In *The stanford encyclopedia of philosophy, ed.* Edward N. Zalta, plato.stanford.edu/entries/technology/. Publisher is: The Metaphysics Research Lab, Center for the Study of Language and Information, Stanford University, Stanford, CA 94305–4115.

Hansson, S.O. 1998. Should we avoid moral dilemmas? *The Journal of Value Inquiry* 32(3): 407–416.

Huff, C. 2008. It is not all straw, but it can catch fire: In defense of impossible ideals in computing. *Science and Engineering Ethics* 14: 241–244.

Hughes, J. 2009. Practical reasoning and engineering. In *Handbook of the philosophy of science. Vol. 9: Philosophy of technology and engineering sciences*, ed. A. Meijers, 375–402. Amsterdam: Elsevier.

Iranzo, V. 1995. Epistemic values in science. *Sorites* 1: 81–95.

Kroes, P. 2009. Foundational issues of engineering design. In *Handbook of the philosophy of science. Vol. 9: Philosophy of technology and engineering sciences*, ed. A. Meijers, 513–541. Amsterdam: Elsevier.

Kroes, P., M. Franssen, and L. Bucciarelli. 2009. Rationality in design. In *Handbook of the philosophy of science. Vol. 9: Philosophy of technology and engineering sciences*, ed. A. Meijers, 565–600. Amsterdam: Elsevier.

Laudan, L. 1984. *Science and values: The aims of science and their role in scientific debate*. Berkeley: University of California Press.

Levi, I. 1986. *Hard choices: Decision making under unresolved conflict*. Cambridge: Cambridge University Press.

Locke, E.A., and G.P. Latham. 1990. *A theory of goal setting and task performance*. Englewood Cliffs: Prentice-Hall.

Locke, E.A., and G.P. Latham. 2002. Building a practically useful theory of goal setting and task motivation: A 35-year odyssey. *American Psychologist* 57(9): 705–717.

Maimon, O., and D. Braha. 1996. On the complexity of the design synthesis problem. *IEEE Transactions on Systems, Man, and Cybernetics – Part A: Systems and Humans* 26(1): 142–151.

McCann, H.J. 1991. Settled objectives and rational constraints. *American Philosophical Quarterly* 28: 25–36. (Reprinted in Mele, A.R. (ed.). 1997. *The philosophy of action*, 204–222. Oxford: Oxford University Press.)

Miller, K.W. 2008. Critiquing a critique. A comment on "A critique of positive responsibility in computing". *Science and Engineering Ethics* 14(2): 245–249.

Nozick, R. 1993. *The nature of rationality*. Princeton: Princeton University Press.

Pahl, G., and W. Breitz. 1996. *Engineering design: A systematic approach*, 2nd ed. Berlin: Springer.

Rosencrantz, H., K. Edvardsson, and S.O. Hansson. 2007. Vision zero – Is it irrational? *Transportation Research Part A: Policy and Practice* 41(6): 559–567.

Schmidtz, D. 1995. *Rational choice and moral agency*. Princeton: Princeton University Press.

Spencer, T., P.M. Guthrie, J. Da Mosto, and C.A. Fletcher. 2005. Introduction: Large-scale engineering solutions to storm surge flooding. In *Flooding and environmental challenges for Venice and its Lagoon: State of knowledge*, ed. C.A. Fletcher and T. Spencer, 241–244. Cambridge: Cambridge University Press.

Stedry, A.C., and E. Kay. 1966. The effects of goal difficulty on performance: A field experiment. *Behavioural Science* 11(6): 459–470.

Stieb, J.A. 2008. A critique of positive responsibility in computing. *Science and Engineering Ethics* 14(2): 219–233.

Part II
Normativity and Artefact Norms

Chapter 7
Valuation of Artefacts and the Normativity of Technology

Sven Ove Hansson

1 Introduction

Neither engineers nor philosophers of technology can evade the old question "Is technology good or bad?" The standard answer is that technology in itself is neither good nor bad. Technology as such is believed to be value neutral, and values only come into play when we use it for good or bad purposes or with good or bad effects. "Ships, and Guns do not fight of themselves, but Men who act and manage them", said William Petty in his *Political Arithmetick* more than 300 years ago.[1]

The question whether technology is good, bad, or neutral, operates in a "macro" perspective, treating technology as a whole. We can also apply value statements to technology in a "micro" perspective, referring to specific technological objects and practices. Such value statements are very common in our everyday communications about technology. As users of technology, we refer to all kinds of technological objects as good or bad, better or worse. Engineers talking about their work do the same. In addition, engineers (but less often everyday users) refer to norms pertaining to technology. Thousands of expert committees around the world are involved in the creation of norms and standards for the construction of technological objects.

---

[1] Petty ([1691] 1899), p. 280. This line of reasoning has been used extensively in the "Guns do not kill" campaigns of weapons lobbies (Haider-Markel and Joslyn 2001). It has also been invoked by other industries producing lethal products. According to Pava and Krausz (1997, p. 343), an argument "similar to the argument put forth by the gun lobby; guns do not kill, people kill", but referring to cigarettes instead of guns "is by far the most important defense for the [tobacco] industry".

S.O. Hansson (✉)
Royal Institute of Technology, Stockholm, Sweden
e-mail: soh@kth.se

M.J. de Vries et al. (eds.), *Norms in Technology*, Philosophy of Engineering and Technology 9, DOI 10.1007/978-94-007-5243-6_7,

How are the macro and micro levels related to each other? Is the presumed value neutrality on the macro level compatible with the extensive use of value statements and normative statements in discussions about technology on the micro level? In order to answer this question, we need to have a closer look at the types of value statements and normative statements that are made about technology.

2 Classifying Value Statements

In order to make this discussion precise enough, we need to have a closer look at the linguistic expressions that we use to express values. There are two major ways to classify these expressions: They can be classified according to what they say about the quantity (amount) of value or according to the standard of evaluation that they refer to.

2.1 Quantitative Classification

Value statements are either comparative (dyadic) or classificatory (monadic). The major comparative value notions are "better", "of equal value", and "at least as good as". The structure of these concepts has been the subject of extensive logical investigations (see Hansson 2001b for an overview). The most important properties that have been investigated are the following two:

Transitivity: If x is at least as good as y, and y is at least as good as z, then x is at least as good as z.
Completeness: Either x is at least as good as y or y is at least as good as x.

Neither of these postulates is uncontroversial; indeed, much of the discussion has been concerned with proposed exceptions to them. In what follows, however, transitivity will often be assumed since it simplifies some of the issues to be discussed.

Natural language contains a long list of expressions that can be used to make classificatory value statements: "good", "best", "very good", "fairly good", "bad", "very bad", etc. In order to clarify their structure, it is useful to relate them to the relation of betterness. We can divide the classificatory value predicates into three major categories: positive, negative, and circumscriptive predicates (Hansson 2001a).

What is better than something good is itself good. Many other value predicates – such as "best", "not worst", "very good", "excellent", "not very bad", and "acceptable" – have the same property. If one of these predicates holds for x, then it also holds for everything that is better than x or equal in value to x. These are the *positive* value predicates.

Similarly, "bad" has the converse property that if x is bad, then whatever is worse than or equal in value to x is also bad. Other predicates that share this property are "very bad", "worst", and "not best". These are the *negative* value predicates.

Ordinary language also contains a third category of value predicates, namely, those that are, intuitively speaking, bounded both upwards and downwards. "Almost worst" and "fairly good" are examples of this category. These are the *circumscriptive* predicates. From a formal point of view, each circumscriptive predicate can be defined as the intersection of one positive and one negative predicate. Thus, "x is almost worst" may be defined as "x is very bad and x is not worst", employing the negative predicate "very bad" and the positive predicate "not worst".

Equivalently, we can define a value predicate H as…

…positive if it is possible for something to be too bad to be H, but it is impossible for something to be too good to be H,
…negative if it is impossible for something to be too bad to be H, but it is possible for something to be too good to be H,
…circumscriptive if it possible for something to be too bad to be H, and it is also possible for something to be too good to be H,

2.2 Classification in Terms of Value Standards

It has often been pointed out that a value statement should be understood as relative to some more or less explicit criterion or standard of evaluation. The criteria commonly referred to provide us with types of values such as instrumental values, aesthetic values, and ethical values (von Wright 1963). Value statements may also be intended to include all aspects, i.e. represent an evaluation that takes everything into account. These have been called "synoptic" values.[2]

We therefore have two groups of value statements. One group consists of the "synoptic" value statements that are *unspecified* with respect to the criteria of evaluation. The second group consists of those that are *viewpoint-specified*, i.e. specified with respect to some criterion of evaluation. Such criteria are often expressed with adverbs such as "morally" and "aesthetically":

"Aesthetically, this is the best of the three designs."
"His behaviour was morally worse than that of his predecessors."
"Economically speaking, this is the best option."

For a more detailed specification of the viewpoint, more extensive phrases than a single adverb may have to be used:

"This is an excellent motor from the viewpoint of fuel economy."

A study of the value statements actually made in various contexts will show that there is also a third group of such statements that has not been much discussed, namely, those value statements that are specified in terms of some category that the object of value belongs to. The following are but a few examples:

[2] Rescher 1968, p. 293. Rawling (1990, p. 495) uses the term "categorical preferences" for essentially the same concept as Rescher's "synoptic preferences".

"She is a good flutist."
"He is a good father."
"This is the worst chainsaw I have ever used."
"The previous version of this computer programme was better than the new one."

I have proposed elsewhere to call this type of statements *category-specified* value statements (Hansson 2006a). They evaluate an object according to the standards for a specified category that it belongs to.[3] Since one and the same object may belong to several such categories, its evaluation may differ according to which category is chosen:

"She is a good engineer but a bad supervisor."
"This is a fairly good kitchen table but it would make very bad desk."

We can therefore divide value statements into three general types with respect to standards of evaluation: unspecified, viewpoint-specified, and category-specified value statements. The lines between them are not perfectly sharp. In particular, the line between unspecified and category-specified value statements is not always easy to draw. Many (arguably all) unspecified value statements are best understood as implicitly category-specified. When we say that a decision is good, we typically mean that it is good *qua* decision.

Category specification and viewpoint specification are distinctly different from each other, and generally speaking, neither can be reduced to the other. Hence, category specification does not limit the viewpoint. When speaking about a "good banker", we typically mean a banker who is good from the various viewpoints that we typically apply to bankers: competence, efficiency, ethical behaviour, etc. But on occasions, viewpoint and category specification are combined into one and the same value statement, as in the following example:

"She is a morally good banker."

In such combinations, viewpoint specification tends to dominate over category specification. Our morally good banker has to be (1) morally good and (2) a banker, but it is not equally obvious that she has to be (3) a good banker. Similarly, as we all know, a meal that is nutritionally excellent need not at all be a good meal, and an aesthetically excellent car need not be an excellent car nor even a good one.

3 Categories of Technological Objects

Category-specified value statements are used in a wide variety of contexts, but it is difficult to find a context in which they are as prevalent as in the evaluation of technological artefacts. A large part of the value statements that we make about technologically produced objects are category-specified. We speak about good cars

---

[3] Note the difference between (1) being a good *x* and (2) being good and being an *x*. See Hansson (2005).

and bad computer programs, and engineers constructing new devices look for better components and materials. This is indeed a characteristic form of value statements in technology (on the micro level referred to above). Therefore, it is unfortunate for the analysis of values in technology that category-specified value statements have not received much attention in philosophical value theory.

Many different types of categories can be used in category-specified value statements. In a general theory of such value statements, a rather extensive typology may have to be developed. In the application of category-specified statements to technological artefacts, a central distinction is that between categories that are defined in terms of functional respectively non-functional characteristics.

Screwdrivers are a functional category. By this is meant that in order to determine whether or not an object is a screwdriver, we have to determine whether or not it has the function of driving screws. Similarly, nutcrackers, pens, airplanes, and CPUs are all functional categories. In fact, most of the objects that we refer to as technological objects (or artefacts) are functionally defined. However, there are also some categories of technological artefacts that are defined according to their physical properties rather than their functions. PVC pipes, copper wires, planks, fibreboards, and ropes are examples of these. In order to determine whether or not an object belongs to one of these categories, we have to find out its physical characteristics rather than its intended function. As these examples indicate, non-functional technological categories are usually raw materials or multipurpose components.

Some categories of technological artefacts are delimited in both physical and functional terms. One example is cogwheels. We would probably not use that designation about a toothed wheel that was constructed for some other purpose than to be connected with another toothed device. Neither would we use it about a wheel that connected with another wheel through some other mechanism than interlinking teeth. Another example is helicopters. A helicopter is a hovering aircraft employing revolving rotor blades. We do not use the term "helicopter" about a hovering aircraft not operating with revolving rotor blades (such as a jet pack or a hovering jet plane). Neither would we use it about a device with revolving rotor blades that are mounted for some other purpose than flying (such as a hypothetical submarine constructed in that way).

However, such combined categories seem to comprise a very small part of the categories we use when describing technological artefacts. For simplicity, they will be left out in what follows.

4 Functional Value Statements in Technology

As I have already mentioned, functional categories tend to dominate in most areas of technological discourse. Consequently, most category-specified value statements made in technological contexts refer to a functional category. In other words, the value-specifying category (such as "screwdriver" in "this is a good screwdriver") usually consists of objects that have a certain function in common.

4.1 Function and Value: A First Approximation

The notion of technological function has been analyzed elsewhere (Vermaas and Houkes 2006; Hansson 2006b; Preston 2009; Vermaas and Garbacz 2009), and I will not go into details here. It should be noted, though, that for some types of technological artefacts the criteria of functionality can be quite complex. The list of your criteria for a good car or a good camera may be quite long, and you are probably prepared to make additions to it based on the experiences you acquire. Most people would not spontaneously include "does not leak toxic substances" on their list of criteria for a good camera, but if they encountered leaking cameras, they would probably add this property to their list. Generally speaking, our list of functional criteria for a type of artefacts is often unfinished and always open to additions due to experiences that we acquire or to technological and social developments.

When a category is specified functionally, this specification will in typical cases also generate value criteria for the category. Hence, a hammer is an object with the function of driving nails or striking blows at material objects. It follows from this that a good hammer is one that satisfies this function well, so that blows can be struck with maximal precision and minimal effort. In general, if we know functionally what an *X* is, then we also know what a good *X* or a bad *X* is. As a first approximation, being a good *X* means to fulfil the defining functions of an *X* to high degree, being a bad *X* means to perform them to a low degree, and similarly for other value terms.

Of course, this rule can only be applied to technological categories that are functionally defined. But in practice, an engineer has no difficulty in understanding what is meant by a "good" steel wire although this concept is not functionally defined: A good steel wire should have the properties required in the various applications where we expect it to be used. In this way, technological artefact types that are not defined in terms of functions are nevertheless evaluated in terms of the functions for which they are typically used. This means that, seemingly without exceptions, the category-specified value statements that we make about technological artefacts refer to the fulfilment of technological function. For functionally defined categories, they refer to fulfilment of the function that defines the category. For categories defined in terms of physical characteristics, they refer to fulfilment of the functions for which objects in the category are typically used.

4.2 Four Types of Categories

In a more precise analysis, we need to take into account that the categories referred to in category-specified value statements are often themselves value-laden. From that point of view, they can be divided into four types.

A category of objects is *unambiguously positive* if the very fact that an object belongs to that category is (generally conceived as) carrying positive value. If I say that a change in a computer program is an "improvement", then I do not need to add that I consider the change to be good. That was already implicit in its classification as an improvement.

Similarly, a category of objects is *unambiguously negative* if by saying that an object belongs to it I am understood to depreciate it. By classifying a part of a computer program as a "bug", I have already made it clear that I see it as a bad part of the program. Similarly, after calling an event an "accident", one does not have to add that it was a bad thing that it happened.

The *value-inert* categories are those for which we have no standard at all. There are no good stars or bad protons. If you have described an object as a star, it would be strange to continue by applying some value to it under that same description. (The object can of course be accessible to valuation with respect to some other category to which it belongs. *Qua* star, the pole star is neither good nor bad. *Qua* navigation point, it is excellent.)

Finally, the *value-ambiguous* categories are those to which both positive and negative values statements can be applied. The term "book" is value ambiguous since by calling something a book, we do not say anything about its value. However, it is not value inert, since we can meaningfully apply value statements to it. This is because we have standards for what it means for a book to be good or bad.

The difference between value-inert and value-ambiguous categories is not "in the nature of things"; instead, it emanates from the human use of things. For a farmer, "cow" is a value-ambiguous category; a good cow is one with a high milk yield and other properties useful in dairy farming. In contrast, "hedgehog" is a value-inert concept. We cannot meaningfully say that a hedgehog is good or bad. We have no value standard for hedgehogs, for the simple reason that we do not use them. If we started to train hedgehogs for circus acts or use these animals in some other way, then we would soon develop a standard for good and bad hedgehogs. (Our ancestors who ate hedgehogs presumably had such a standard.)

As this example shows, categories denoting things that we use are not value inert. It is therefore no surprise that there do not seem to be any value-inert categories referring to technological artefacts. In the valuation of technology, we can focus on the other three types of categories, those that are unambiguously positive, unambiguously negative, or value ambiguous.

4.3 Asymmetries in the Use of Value Terms

The unambiguously negative categories are those for which we have a standard or "ideal type" that we never want to see realized. These are categories to which we do not normally apply positive value terms. A person who is proficient at killing other persons would not be called a "good murderer". Neither would we talk about a "good murder weapon" or a "good landmine". Our unwillingness to use such phrases depends, of course, on our moral values, not on conceptual difficulties. We have no difficulty in understanding what someone with a different value system such as a hitman or a war criminal would mean if they used these phrases.[4] Similarly, if phrases like "a good accident" or "a good mistake" are heard, then there are strong reasons to suspect equivocation.

---

[4] It is important to distinguish here between the technical object or system as a whole and its various parts. Even though we would not talk about a "good torture rack", we can say that "a good screw" or "a rope of high quality" was used in its construction. These objects are then treated in isolation and judged according to general criteria for their respective category.

However, it is only positive evaluations that are excluded. Negative evaluations are fairly unproblematic; we may for instance speak of the worst murderer in the country or a really bad accident. As these examples show, when we refer to an unambiguously negative category, we use adjectives on the negative side of the value scale to express that an object fulfils the function typical of that category to a high degree. The worst murderer in the country satisfies the criteria for being a murderer to a very high degree. (In contrast, the worst cook, or mayor, or policeman, satisfies the standard for their respective category to the lowest degree.) Similarly, bad mistakes and bad accidents satisfy the criteria for mistakes respectively accidents to a very high degree. This can be taken as a litmus test for a term to be unambiguously value negative (Hansson 2006a).

Since a bad mistake is one that satisfies the criteria for a mistake to a high degree, and we avoid talking about good mistakes, how can we at all express that a mistake satisfies the criteria for its kind only to a small degree? There are no value terms left, and therefore, this is expressed with non-valuational terms; we refer in such cases to a "small mistake". Similarly, a school shooter who does not manage to kill as many people as most other school shooters would be referred to as "less successful" than the others. We would not call him "better" than the others (and certainly not "worse").[5]

Since the use of positive value terms is avoided for unambiguously negative categories, we could perhaps expect to find a similar tendency to avoid negative value terms for unambiguously positive categories. But there is no such tendency. We have no problems in applying either positive or negative terms to unambiguously positive value terms. We can speak about bad as well as good antidotes, life jackets, and defibrillators, whereas we prefer not to apply the word "good" to a poison, a torturer's drowning pit, or an electrocution trap for humans.

Value-ambiguous categories can, unsurprisingly, have both positive and negative value terms applied to them. We can talk about a good or a bad bolt, and the same applies to copper wires, planks, electric motors, and other value-ambiguous technological artefacts. In this respect, value-ambiguous categories – that include the vast majority of the categories applied to technological artefacts – are treated in the same way as unambiguously positive ones. In both cases, we use a positive value term to express that an object fulfils its function to a high degree. Hence, a good screw is one that satisfies the functions of a screw to a high degree. This applies irrespectively of whether the screw is used in a landmine or a pacemaker. Similarly, if an electric motor fulfils the functions of a motor to a high degree, then it is a good motor, irrespectively of whether it is used in a torture rack or a medical ventilator.

The findings of this section are summarized in Table 7.1. It reveals a simple but important pattern: Value terms are applied to value-ambiguous categories in the same way as they are applied to unambiguously positive categories, whereas their application to unambiguously negative categories is different. This has important implications for our evaluations of technological artefacts: When a technological

[5] On the other hand, we would not hesitate to call the more efficiently killing school shooters "worse" than the others. This is an example of the asymmetry between betterness and worseness that has been pointed out by Brogan (1919) and others.

Table 7.1 The use of value terms to express fulfilment of functional categories in the different types of categories identified in the text

	Value terms used to denote that the defining function is satisfied…	
Type of category	…to a high degree	…to a low degree
Unambiguously positive	Positive terms	Negative terms
Unambiguously negative	Negative terms	(Non-value terms)
Value ambiguous	Positive terms	Negative terms
Value inert	(Undefined)	(Undefined)

artefact can be used for both good and bad purposes, we talk about it in the same way as we would talk about an artefact that is used only for good purposes, not in the way we would talk about an artefact used only for bad purposes. A bolt is "good" if it satisfies the (functional) criteria for a bolt to a high degree, irrespectively of the purposes for which we use it. This linguistic practice can be described as a *presumption of good use*. We will return to its implications in Sect. 6.

5 Norms

The distinction between facts and values is one of the most influential philosophical insights. The importance of not confusing the two is emphasized far outside of philosophy. In contrast, the distinction between norms and values is seldom discussed. Even professional philosophers often take it for granted that the fact–value and is–ought distinctions are essentially the same. Of course, "is" refers to facts, but "ought" is not about values. It is a normative concept, not a value concept.

Normative concepts differ from value concepts in being directly action guiding. If I say that in my view, all things considered, you ought to do *X*, and then continue by saying that I do not recommend you to do *X*, then I will be considered to behave inconsistently. In contrast, I may consistently say that doing *X* would be a good action (or the best action, an action that it is bad to refrain from, etc.) and yet consistently add that I do not recommend you to perform it. One possible reason for this may be that *X* is supererogatory (beyond the call of duty) (Hansson 1991).

This dissimilarity prevents norm statements and value statements from being synonymous or interdefinable. However, it does not follow that they cannot be extensionally equivalent (Moore 1912, pp. 172–173). (Two monadic predicates T_1 and T_2 are extensionally equivalent if and only if for all arguments x, $T_1(x)$ holds if and only if $T_2(x)$ holds. They are intensionally equivalent, i.e. synonymous, if and only if for all arguments x, the statement that $T_1(x)$ holds has the same meaning as the statement that $T_2(x)$ holds. I have argued elsewhere that the conditions of rational assent for value concepts and normative concepts are extensionally connected and that this connection can be modelled with the tools of logic (Hansson 1991). This means, more precisely, that a set of value statements and a set of normative concepts may be incompatible even though each of the two sets, taken alone, is consistent.

Norms have important roles in technology. Engineers make extensive use of explicit norms in written codes and standards that provide detailed specifications for a wide variety of technological products, practices, and procedures. In addition, engineering practice is largely ruled by non-codified norms. Engineers designing a new product work with a more or less explicit list of design specifications or goals. The ideal outcome of the design process would be a product that satisfies all these specifications. In practice, that seldom happens. Compromises have to be made in the course of the design process.

Most of the codified norms are *standard-setting norms*. By this, I mean that they express standards that have to be fully satisfied. They have the same relation to values as most moral and legal norms: You have to comply (completely) with the norm, and if you do not, then that is bad. Many of the uncodified norms have another role. They are *ideal-setting norms*, expressing ideals to which the engineer is supposed to strive. It is not expected that one will be able to comply completely with the norm, but the outcome is better, the closer one manages to come to it. Codified norms tend to be standard setting, and non-codified norms to be ideal setting. However, the correlation is not perfect. In particular, some non-codified norms, including many safety norms, are standards rather than ideals.

Ideal-setting norms are closely related to category-specified values. A good screwdriver, for instance, is a screwdriver that comes reasonably close to the norms for an ideal screwdriver. More generally, let *X* be a category that is either value-ambiguous or unambiguously positive. If there is an ideal for *X*, then an object is a good *X* to the extent that it comes close to that ideal.

6 Conclusion

Finally, let us return to the question that we started with: Can the standard view that technology as such is value neutral be reconciled with the extensive use of value statements in discussions on technology on the micro level?

The presumption of good use that was identified at the end of Sect. 4 has an essential role in answering that question. This presumption expresses a deeply rooted linguistic habit: When a technological artefact can be used for both good and bad purposes, we call it "good" if it satisfies the (functional) criteria for its category to a high degree, without paying attention to whether we use it for good or bad purposes. This means that in terms of micro level valuations, we treat technology as good unless there are reasons to do otherwise.

The presumption of good use is not easily reconcilable with the precept that technology as such should be treated as value neutral. The patterns of thought encoded in our linguistic practices appear to be more positive towards technology than what the tenet of value neutrality enjoins us to be. There seems to be a real tension between the presumed value neutrality of technology and the ways in which we apply values and norms to technological artefacts. Our linguistic practices are more easily compatible with a positive view of technology. According to that view, it is a good thing if we manage to produce general-purpose technological artefacts that increase our capabilities by satisfying the various functional criteria that we establish for such devices: better steel, better electric motors, better bolts, etc. These artefacts can be used either for good or bad purposes, but according to this view, the extension of human capabilities through technology is in itself a phenomenon with a positive value, rather than being value neutral. To meet this view, the usual motto "technology is neutral but it can be used for good or bad purposes" would have to be replaced by something like "most technology is good but there is also bad technology, and good technology can be used for bad purposes".

It is not my purpose to argue here in favour of this or any other view of technology on the macro level. Obviously, our linguistic practices should not dictate our moral standpoints. Instead, I will offer two much less far-reaching conclusions. Firstly, there are connections between the macro and micro levels in the evaluation of technology, and we need to explore these connections. Secondly, our current linguistic practices encode a positive view of technology, and this has to be taken into account in the articulation of other views on the values of technology.

Appendix: The Logic of Category-Specified Value

The categories that we use to specify values can be more or less general. A compass saw is a saw, and a saw is a tool. A Mazda is a car, and a car is a motor vehicle. A heart surgeon is a surgeon, and a surgeon is a physician. An amateur flutist is a flutist, and a flutist is a musician. How do our evaluations in subcategories relate to our evaluations in the more general categories in which they are included? Is a good compass saw a good saw? Is a good amateur flutist a good flutist? This is an issue of a type best elucidated with formal logic. What follows is a first outline of how a formal logic of category-specified values can be constructed.

Categories and Their Elements

To begin with, it is important to note that we are dealing with categories, not sets. Categories are intentionally defined. For each category C, there is a unique set that consists of all members of that category. This set can be denoted |C|. Hence, if C is the category of chainsaws, then |C| is the set of all chainsaws. (There may be non-intentional ways to specify |C|, such as listing all its elements.)

Valuations will refer to objects included in the categories, not to the categories themselves. I will use the symbol ε to denote the relationship of belonging to a category, whereas ∈ denotes set membership as usual. If x is an actual (non-hypothetical) object, then

$x \varepsilon \mathrm{C}$ if and only if $x \in |\mathrm{C}|$.

Subcategories

A *subcategory* of a category C is a category C′ such that any (actual or possible) object satisfying the criteria for inclusion in C′ must also satisfy the inclusion criteria for C. I will use the symbol $\underline{\ll}$ to denote the subcategory relationship; hence, $C' \underline{\ll} C$ denotes that C′ is a subcategory of C.

There are many ways to construct subcategories of, for instance, the category "saw". We have the subcategories "saw made in Germany", "cheap saw", etc. However, since "saw" is a functional category, a particularly interesting type of subcategories are those that specify the function more precisely than the original category. Given a functional definition of a saw, we can define "compass saw", "chainsaw", etc., by adding further specifications of its function. These are *functional subcategories*. In the same way, "surgeon" is a functional subcategory of "physician", and "flutist" a functional subcategory of "musician".

I will use the symbol $\underline{\ll} f$ to denote functional subcategories; hence, $C' \underline{\ll} fC$ denotes that C′ is a functional subcategory of C.

The following are two plausible properties of subcategories:

If $A \underline{\ll} fB$, then $A \underline{\ll} B$.
If $A \underline{\ll} B$, then $|A| \subseteq |B|$.

The converse properties do not hold, i.e. $A \underline{\ll} B$ does not in general imply $A \underline{\ll} fB$, and $|A| \subseteq |B|$ does not in general imply $A \underline{\ll} B$.

Value Predicates

In standard preference logic, there are three dyadic operators, ≥ (at least as good as), > (better than), and ≈ (equally good as). Commonly, the second and third of these are taken to be definable in terms of the first; thus,

$x > y$ if and only if $x \geq y$ and *not* $y \geq x$,
$x \approx y$ if and only if $x \geq y$ and $y \geq x$.

$x \geq y$ is read "*x* is at least as good as *y*". In order to make it category-specified, we have to add the category to the relation; hence,

$x \geq Cy$ "*x* is at least as good C as *y*"

and similarly
$x > Cy$ "*x* is a better C than *y*",
$x \approx Cy$ "*x* is an equally good C as *y*".

The relations ≥C, >C, and ≈C are only applicable within C. In other words, $x \geq Cy$, $x > Cy$, and $x \approx Cy$ are all undefined unless both $x \in C$ and $y \in C$.

The most important monadic value terms are G ("good") and B ("bad"). To make them category-specified, we can provide them with categories as subscript in the same way as the dyadic predicates. Hence,

$G_C x$ “x is a good C”,
$B_C x$ “x is a bad C”.

These predicates only apply in the domain indicated by their index; hence, $G_C x$ and $B_C x$ are both undefined unless $x \varepsilon C$.

The following are plausible postulates for the monadic predicates:

Positivity of “good”: If $G_C x$ and $y \geq_C x$, then $G_C y$.
Negativity of “bad”: If $B_C x$ and $x \geq_C y$, then $B_C y$.
Mutual exclusiveness: $G_C x$ and $B_C x$ do not both hold.

Some Valid Inference Principles

The following inference pattern seems to hold for functional subcategories:

x is a good A
A is a functional subcategory of B
∴ x is a good B

More succinctly
if $G_A x$ and $A \underline{\ll}_f B$, then $G_B x$.

For concreteness, let A denote “compass saw” and let B denote “saw”, or let them denote “zoologist” respectively “scientist”, or (contrary to prejudice) “violist” respectively “string player”.

The corresponding relationship does not hold in general for non-functional subcategories, i.e. it does not hold in general if we replace «f by «. Unfortunately, a good cheap camera is seldom a good camera. Furthermore, the corresponding property does not hold at the other end of the value scale, i.e. the following property does *not* hold in general:

If $B_A x$ and $A \underline{\ll} f B$, then $B_B x$.

A bad piccolo player need not be a bad flutist. A multipurpose saw may be a bad hacksaw but nevertheless a good saw since it is a good compass saw. The logical asymmetry between the positive and negative ends of the value scale that has been noted in other contexts (Hansson 2001a, p. 147) seems to be prominent in the logic of category-specified value statements.

However, the following reverse property seems to hold at the negative end of the value scale:

If $B_A x$, $x \varepsilon B$, and $B \underline{\ll} f A$, then $B_B x$.
It can also be expressed in syllogistic form:
x is a bad A
x is a B
B is a functional subcategory of A
∴ x is a bad B

To exemplify this, if *x* is a flutist and a bad musician, then we can conclude that *x* is a bad flutist. This property has a close analogue in social choice theory. As can easily be verified, it is in fact nothing else than Sen's property α (also called "Chernoff"), with $\subseteq$ replaced by $\ll$f and "chosen" by "bad".[6] However, as can also easily be verified, this property does not hold in general for functional subcategories at the positive end of the value scale (which is where it is applied in social choice theory). A good musician can be a pianist without being a good pianist. (She may for instance be a good violinist who doubles with less success on the piano.)

In an interesting paper, Maarten Franssen (2010) points out that there are cases in which the α property holds at the positive end of the value scale. The example he mentions can be expressed syllogistically as follows:

x is a good musical instrument
x is a saxophone
"Saxophone" is a functional subcategory of "musical instrument"
∴ *x* is a good saxophone

The reason why the pattern works in this particular case seems to be that "saxophone" is a non-convoluted functional subcategory of "musical instrument", in the following sense:

> A is a *non-convoluted* functional subcategory of B if and only if (1) A is a functional subcategory of B, and (2) it holds for all functional subcategories A' of B that either A $\underline{\ll}$f A', A' $\underline{\ll}$fA, or A and A' are disjoint.

The definition can be illustrated with Franssen's example. Let A be the category of saxophones and B that of musical instruments. Then it holds for any other instrument category A' that either (1) A $\underline{\ll}$fA', as if A' denotes woodwind instruments, (2) A' $\underline{\ll}$fA, as if A' denotes soprano saxophones, or (3) A and A' are disjoint, as if A' denotes timpani. There is no instrument category that stands in the same relation to saxophones as knives to weapons. (Some knives are weapons but some are not, and there are weapons that are not knives.) The following syllogism is suggested by Franssen's example:

x is a good A
x is a B, and B is a non-convoluted functional subcategory of A
∴ *x* is a good B

Incidentally, not all categories of musical instruments are non-convoluted. Five-stringed instruments are available that can serve both as a violin and a viola. (They are usually called five-stringed violins.) It can be appropriate to say about one of these instruments that it is a good string instrument (since it is a good viola) but a bad violin. Since "violin" is not a non-convoluted category if we take the five-stringed instruments into account, this example does not contradict the proposed syllogism.

---

[6] For an excellent introduction to properties referred to in social choice theory, see Moulin (1985).

References

Brogan, Albert P. 1919. The Fundamental Value Universal. *Journal of Philosophy, Psychology, and Scientific Methods* 16: 96–104.

Franssen, Maarten. 2010. The goodness and kindhood of artefacts. In Norms in technology (Philosophy of engineering and technology), ed. Marc J. de Vries, Sven Ove Hansson, Anthonie W.M. Meijers, pp. 155–169. Springer, Dordrecht.

Haider-Markel, Donald P., and Mark R. Joslyn. 2001. Gun policy, opinion, tragedy, and blame attribution: The conditional influence of issue. *Journal of Politics* 63: 520–543.

Hansson, Sven Ove. 1991. Norms and values. *Critica* 23(67): 3–13.

Hansson, Sven Ove. 2001a. *The structure of values and norms*. Cambridge: Cambridge University Press.

Hansson, Sven Ove. 2001b. Preference Logic. In *Handbook of philosophical logic*, 2nd edn., ed. Dov Gabbay and F. Guenthner, 4: 319–393. Dordrecht: Kluwer Academic Publishers.

Hansson, Sven Ove. 2005. Aesthetic functionalism. *Contemporary Aesthetics*, 3. Full text: http://www.contempaesthetics.org/newvolume/pages/journal.php.

Hansson, Sven Ove. 2006a. Category-specified value statements. *Synthese* 148(425–432): 2006.

Hansson, Sven Ove. 2006b. Defining technical function. *Studies in History and Philosophy of Science* 37(19–22): 2006.

Moore, G.E. 1912. *Ethics*. London: Oxford University Press.

Moulin, H. 1985. Choice functions over a finite set: A summary. *Social Choice and Welfare* 2: 147–160.

Pava, Moses L., and Joshua Krausz. 1997. Criteria for evaluating the legitimacy of corporate social responsibility. *Journal of Business Ethics* 16: 337–347.

Petty, William. [1691] 1899. *Political Arithmetick Or a Discourse Concerning The Extent and Value of Lands*. (Reprinted in Charles Henry Hull (ed.) *The economic writings of Sir William Petty: Together with the Observations upon the bills of mortality more probably by John Graunt*, Cambridge.)

Preston, Beth. 2009. Philosophical theories of artifact function. In *Handbook of the philosophy of science, Vol. 9: Philosophy of technology and engineering sciences*, ed. Anthonie Meijers, 213–233. Amsterdam: Elsevier.

Rawling, P. 1990. The ranking of preference. *Philosophical Quarterly* 40: 495–501.

Rescher, N. 1968. *Topics in philosophical logic*. Dordrecht: Reidel.

Vermaas, P.E., and P. Garbacz. 2009. Functional decomposition and mereology in engineering. In *Handbook of the philosophy of science, Vol. 9: Philosophy of technology and engineering sciences*, ed. Anthonie Meijers, 235–271. Amsterdam: Elsevier.

Vermaas, P.E., and W. Houkes. 2006. Technical functions: A drawbridge between the intentional and structural natures of technical artefacts. *Studies in History and Philosophy of Science Part A* 37: 5–18.

von Wright, G.H. 1963. *Varieties of goodness*. London: Routledge & Kegan Paul.

Chapter 8
Artefactual Norms

Krist Vaesen

1 Introduction

There is a long-standing tradition of characterizing artefacts in terms of purposes. Artefacts are physical entities, subject to the laws of nature, but what makes them different from stones and clouds—that is, what makes them artefacts—is the fact that they are intentionally produced and used by human beings to realize certain goals (see, e.g., Simon 1969; Hilpinen 2004; Kroes and Meijers 2006; or much earlier, see Aristotle's Physica). Consequently, an evaluation of an artefact is typically cast in terms of, what von Wright (1963a) calls, its instrumental goodness. An artefact is good if it indeed has the capacity to produce a certain practical end, bad otherwise.

One obvious sense one could give to instrumental goodness is the following: the ability to realize a proper function. That is, when assessing whether an artefact is good, one focuses on one particular practical end, namely, the instrumental utility associated with the artefact's function. A stapler, for instance, is for fastening stacks of paper together. A good stapler, then, is just a stapler that can perform that particular function.

The notion of function plays a normative role here: it tells us what an artefact ought to do (see, for instance, Hardcastle 2002; Franssen 2006; Chap. 9 by Hughes, this volume). The function of an artefact is an evaluative standard against which we can set the artefact's performance (Neander 2004; Vaesen 2006, 2008, see below). If the artefact fails to meet that standard, we have reason to question its goodness.

While functional considerations cover a substantial part of what makes for good artefacts, artefacts typically serve purposes other than functional and hence can be evaluated for serving these purposes (just as) well. Consider, for instance, the two

K. Vaesen (✉)
Eindhoven University of Technology, Eindhoven, the Netherlands
e-mail: k.vaesen@tue.nl

M.J. de Vries et al. (eds.), *Norms in Technology*, Philosophy of Engineering and Technology 9, DOI 10.1007/978-94-007-5243-6_8,

software packages MS Word and its free open source cousin OpenOffice Writer. The packages, let us assume, are equally good at performing the very same function (i.e., word processing). Nonetheless, some think OpenOffice Writer is good (and MS Word bad), simply because OpenOffice Writer conforms to open source standards and as such allows users to adjust the code to their own preferences. Some other examples: a stapler might be effective for fastening paper together but perform poorly in terms of ease of manufacture; a car might be good at transportation but score low on safety; a lamp might be good at giving light, though bad in terms of environmental impact; a sheet of paper might be perfect to write on but useless for printing if not dimensioned conform to DIN standards; and so forth.

Just like the notion of function, these kinds of considerations relate to human purposiveness and as such involve a similar form of normativity. The DIN standard, for example, determines how sheets of papers ought to be like. Any sheet of paper failing to meet that standard is, at least when the goal concerns printing, a suboptimal sheet of paper.

In comparison to its functional counterpart, this kind of nonfunctional artefactual normativity has received little attention in the philosophical literature on artefacts.[1] In this chapter, I try to show why this is a mistake. I will give an overview of the different kinds of nonfunctional norms involved in both artefact use (Sect. 3) and artefact design (Sect. 4), and argue that without these, any characterization of (present-day) artefacts remains wanting (Sect. 5). But before I do so, I need to make precise what I mean when I talk about norms.

2 What's in a Norm?

The standard dimensions of a piece of A4-paper are 210 mm × 297 mm. So wherever you buy A4-paper, you ought to receive the same-sized kind of thing. In other words, there is a standard or norm any piece of A4-paper ought to conform to.[2]

Gasoline produced by Shell, Texaco, or Total has a nearly identical chemical composition. It ought to work for any brand of car; it ought to be identical in Germany, Switzerland, and France. It is not that oil companies are mandated by law to produce such a standardized type of gasoline; nevertheless, it is in their own

[1] In contrast, these considerations are ubiquitous in industries concerned with quality management. For three particularly instructive case studies, see Bucciarelli 1996. These studies (of the development of x-ray scanners, photoprint machines, and photovoltaic energy system, respectively) forcefully demonstrate the many nonfunctional considerations entering the design process, and how they are the subject of intense negotiations.

[2] In fact, the term standard seems more accepted in English-speaking countries, whereas in Germany, France, and Holland, one does not refer to the ISO 9001 standard but respectively to the ISO 9001 "Norm," "norme," and "norm." In these countries, moreover, the term normalization is preferred over the term standardization—as in "Deutsches Institut fur Normung," "Association Francaise de la Normalisation," and "Nederlands Instituut voor Normalisatie."

interest to do so. If Shell would decide to depart from the norm, customers would go Texaco. Gasoline ought to be such and such; it ought to have this-and-that chemical composition.

So it looks like there exist product norms regarding paper and car fuel—and credit cards, lamp fittings, voltages, transmission signals, Big Macs, and so forth. At least, these common sense, natural language examples are in favor of such a conclusion.[3]

In philosophical accounts, however, norms are typically reserved to actions. According to G.H. von Wright, for instance, norms prescribe actions to agents and take the general form "if you want q, you must do p" (von Wright 1963b; for similar agent-centered accounts of normativity, see, e.g., Raz 1975; Dancy 2004). If so, there seems little room for norms on artefacts (or any other object, for that matter). Indeed, there is no obvious way of fitting an artefact in von Wright's general normative scheme.

One could try to resolve the conflict between natural language and philosophical accounts of normativity by arguing that the product norms engineers talk of can be reduced to norms on actions. A safety norm for cars, for instance, can function as a motivational element in human behavior. In von Wright's scheme: "If you want to design a car, you must make a car that satisfies safety norm X." Safety norm X, then, is normative in the sense that it tells one what to do, when designing a car, namely, to design a safe car.

The main problem with this line of reasoning is that safety norm X does not so much tell us what to do as what to aim for. Consequently, norm X can still be mapped onto a manifold of agentive norms; there is no a priori entailment from a specification of how a car ought to be like to a specification of engineering behavior. Car safety regulations, thus, do typically not concern procedures (actions) but end-products (artefacts). Car manufacturers are free to develop their own favored production method, as long as the produced goods satisfy product regulations. A product norm, thus, might exist without there being any clear recommendation for action linked to it.

Elsewhere (Vaesen 2008) I made the following suggestion to accommodate this fact. I defined a norm as a human-defined and institutionalized measure, a comparative standard of how things ought to be or ought to be done. For example, in order for a human agent to meet the norms of rationality, he ought to act so and so; in order for an artefact to meet norms of quality, it ought to be so and so or ought to perform this and that. On this account, then, the difference between agent and artefactual normativity would be related to what the norms are about: in agent normativity, norms are about human beings and their actions, artefactual normativity on the other hand concerns artefacts.

[3] In the ISO definition of a standard, norms applying to artefacts are mentioned as well (see http://www.iso.org): "Document, established by consensus and approved by a recognized body, that provides for common and repeated use, rules, guidelines, or characteristics for activities or their *results* aimed at the achievement of the optimum degree of order in a given context." [italics mine]

Norms conceived in this way are relevantly similar to the descriptive norms that, as Neander (2004) remarks, are used in some other contexts: for example, in talk of statistical norms and in talk of normal functioning in biology. Considering the normativity of biological functions, the author puts it like this: "To attribute a natural function … to something is to attribute an evaluative standard to it that it could fail to meet, even chronically […]." On this account, a function is a comparative measure to which we refer when the object in question is malfunctioning; in case an item is not functioning as it is (standardly) supposed to function, it is malfunctioning.

An important aspect of norms, on my account, is the fact that they are institutionalized (as opposed to private). It means that the application of a norm as a comparative standard is a collective pattern of action that is socially enforced and has some measure of stability. For instance, ISO norms are developed and used by a collective of agents to evaluate products and product methods. Furthermore, they are socially enforced: some by law, others by virtue of the authority of ISO, still others simply within the manufacturer-customer relationship. In the latter case, customer groups for instance may stop buying products not conforming to the ISO norm, thereby forcing "non-obedient" manufacturers to comply.

Finally, I need to make clear how such norms operate in normative judgments in which the goodness of an artefact is evaluated. To say that a car is good means it complies to present-day car standards. Such standards are typically manifold; they relate obviously to the car's ability to perform its function but also to its safety, its environmental impact, its ease of use, and so forth. If it doesn't comply to these norms, we have reason to call it inferior. In the following two sections, I will give a systematic overview of the kinds of norms artefacts may (need to) comply to.

3 Artefact Use and Norms

As said, the standard interpretation of an artefact's goodness is instrumental. An artefact is good if acting with it realizes a desired goal—in particular, the goal correlated with the artefact's proper function. The kind of instrumental reasoning behind this construal can be captured in the following simple means-end scheme:

[1]
S wants to bring about A.
Using artefact x for its functional goal A brings about A.
Therefore, S uses artefact x for its functional goal A.

In this sort of characterization of artefact usage, artefact usage is simply considered as a means to an end A. Importantly, it is compelling to the extent that artefact x indeed has the capacity to realize its functional goal A—that is, to the extent that x functions properly. According to this simple means-end relation, then, it is quite reasonable to equate instrumental goodness with functionality.

I believe, however, that [1] is a poor characterization of artefact usage and, consequently, that it yields a poor characterization of artefactual goodness. A better characterization should accommodate at least three additional facts: [i] that S typically can choose between a set of alternatives x0, x1, x2, ..., xn to bring about A; [ii] that S's usage of xi may yield undesirable side effects; [iii] and that usage of artefact xi is effective only if used in conjunction with a set of other artefacts (y1, y2, ... ym). Conditions [i] and [ii] are in fact quite common in accounts of means-end reasoning (see, e.g., Bratman 1987; Walton 1990); if S can choose between options other than x, or if the execution of x conflicts with the other ends S pursues, the inference to x, even if conducive to A, may be annihilated. Condition [iii] is more specific to artefact usage; it reflects the fact that many artefacts are useless if used in isolation. To use a simple electric device as a hairdryer, for instance, one not only needs to use the hairdryer itself but also a socket in the wall. The functionality of the hairdryer depends on the functionality of the socket. If one's aim is to dry one's hair, one has compelling reason to use a hairdryer on the condition that the hairdryer's plug fits the socket in the wall.

In sum, a more adequate representation of the means-end relation involved in artefact usage is the following:

S wants to bring about A.
Using artefact x1, x2, ... or xn in conjunction with artefacts (y1,y2, ..., ym) will bring about A. [2]
Using artefact xi in conjunction with artefacts (y1, y2, ... or ym) is more acceptable than not bringing about A.
Therefore, S uses artefact xi in conjunction with artefacts (y1, y2, ... or ym) to bring about A.

In this scheme, the second clause captures conditions [i] and [iii]. It states that, given the availability of complementary artefacts (y1, y2, ... or ym), S may choose to operate any of the functionally equivalent artefacts x1, x2, ... or xn. The third clause relates to condition [ii]: the side effects produced by operating any of the alternative artefacts (or artefact sets) shouldn't cancel out the desirability of A.

Conditions [i]–[iii] in fact fit with a common distinction in the literature on standardization: standards can be subdivided in quality standards (condition [i]), compatibility standards (condition [ii]), and interference (condition [iii]) standards (see, for instance, Blind 2004; de Vries 1999). Let me explain compatibility and interference first, and then turn to quality.

3.1 Compatibility

The most straightforward class of artefactual norms are those related to compatibility. Compatibility (obviously) has to do with an artefact' s links to other artefacts. Only if xi is compatible with the other available resources (y1, y2, ... or ym), S will be able to bring about xi's functional goal A. Although perfectly fit for its functional goal, a well-functioning hairdryer that does not conform to local plug standards is simply deadweight. Likewise, we expect a USB stick to fit in any USB port, and conversely, we expect any computer with a USB port to be able to read what's on the stick, any decent standard web browser should be able to open any standard web page, any cellular phone should be able to receive and send signals to commonly used mobile networks, DIN-format sheets of paper should be compatible with DIN-format envelopes, and so forth.

All these examples involve nonfunctional normativity. A plug standard, for instance, tells us how a plug ought to be, irrespective of—or over and above—the electric device being able to perform its function. The standard is a comparative and stable measure that is socially enforced, for instance, through the work of national and international institutes of standardization. A good hairdryer, then, is one that is able to perform its function and can do so because it is compatible with local auxiliary devices.

One obvious reason why we value compatibility is the fact that we, as economic beings, want to reduce so-called switching costs (Farrell and Shapiro 1988; von Weizsäcker 1982). We hope the things we buy are compatible with the devices we already have, so that we do not need to replace them as well. For instance, each DVD has a region code, specifying the area of the world in which playback is intended. A DVD player, now, will only play discs that contain its region code. So if the region codes on regular DVDs would change on a monthly basis, most people would stop buying them, since to play them, they would each time need a new DVD player as well. Furthermore, compatibility standards make the exchange of artefacts much simpler, as such extending their applicability. It is a quality of my USB flash drive, for example, that I can plug it in my home computer as easily as in my office machine and in the machine of my neighbors.

3.2 *Interference*

Norms of interference regulate the extent to which an artefact may affect other entities—people, other artefacts, or the environment. As an example of the first, tanning beds are harmful to the human eye and skin, so that they should be used with care. Second, concerning the environment, while performing their function effectively, cars are a significant source of air pollution. Finally, broadcasting devices often disturb the working of other artefacts, such as electric appliances.

These examples can be framed in terms of the means-end inference [2], that is, in terms of conflicts between the various goals that average customers pursue (or in some cases, such as in regulations concerning the environment, must pursue). If, for instance, our goal is to make a phone call, we will balance the desirability of that goal with the desirability of other goals, say, our health. If using a particular phone to make a phone call would incur, say, an unacceptable risk of cancer, one might refrain from using that carcinogenic phone. So a telephone is good to the extent that it, while functioning properly, does not interfere too much with the other things we cherish.

At this point, one might argue that norms of interference are, at bottom, just functional norms; the function of a cellular phone is not just to allow phone calls but to allow phone calls without inflicting any disease. This kind of objection is problematic in two respects. First of all, function ascriptions typically do not contain such level of detail. The cars produced in the 1950s, although not being as safe as contemporary cars, are still cars; not any advance in automotive technology is followed by a redefinition of the function of the products that the automotive industry produces (viz., cars). If not, the cars of the 1950s would have a different function than present-day cars—transportation vs. safe transportation—making it particularly difficult to compare the two types of cars, for functional equivalence, it seems, is a necessary condition to assess the instrumental goodness of two artefacts.

Second, returning to our cellular phone example, apart from not causing any disease, there is a limitless amount of things a mobile phone should not do: it should not make the noise of an airplane, it should not electrocute its user, it should not teleport the things it is in contact with, and so forth. Such infinite conjunctions of don'ts would make functional descriptions impractical, and worse, highly uninformative. My suggestion, thus, is not to inflate the notion of function and keep functional norms separate from norms related to interference.

3.3 Quality

In many cases, subjects face an embarrassing variety of choice to pursue their goals. For instance, if S wants to prepare an electronic document, she would have several options: she might use Microsoft Word or OpenOffice Writer or some LaTeX suite or AbiWord or WordPerfect or StarOffice Writer or Notepad or Wordpad or NoteTab or Gedit, and so on. On pain of not realizing her goal, S needs to make a choice. Supposing that the alternatives are equal with respect to compatibility and interference, on what other basis could S make a choice? Put differently, if S faces several equivalent (with respect to function, compatibility, and interference) artefacts x1, x2, ... or xn, what artefactual difference maker would make S choose xi rather than xj?

Several considerations may play a role here, all of which I will subsume under the heading "quality." A first consideration is quite obvious: some artefacts are simply better in performing a certain function than others. Microsoft Word and OpenOffice Writer and other packages might have the same function but differ with respect to some performance criterion. Word or Writer might be more reliable (e.g., crash less), for example, or consume less computational resources. Hence, the goodness of either package would be determined not just in terms of the software's capacity to assist in document preparation (for that is a capacity both packages have) but, relatively, in terms of how effectively/reliably they are at doing so (for a detailed account of effectiveness and reliability, see Chap. 9 by Hughes, this volume).

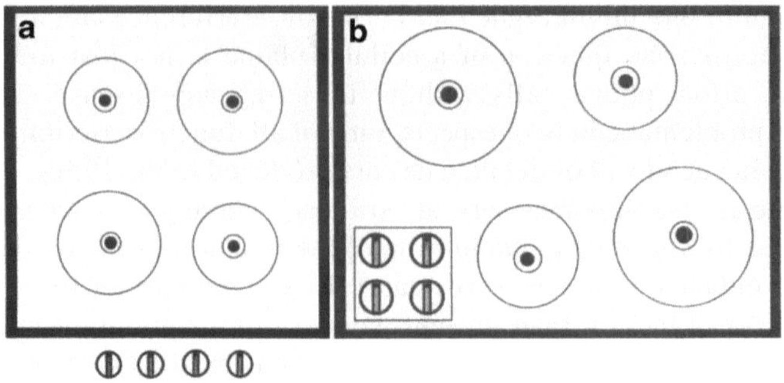

Fig. 8.1 (**a**) Arbitrary arrangement of stove controls; (**b**) intelligible arrangement of controls (Adapted from Norman 1988)

A second consideration has to do with usability. Consider, for instance, an example of Norman (1988). It is about the arrangement of burners and controls on kitchen stoves. Most of them have four burners, arranged in a rectangle, each being operated by an individual control. Some of them look like the ones depicted in Fig. 8.1a. The arrangement in 1a is arbitrary: it is hard to tell which control goes with which burner. With some trial and error, one will probably find out, but next time one uses the thing, one will have to start trying and erring again—unless one manages to memorize the correct arrangement.

For designers, there is an easy solution to the problem, however. They only need to make use of a spatial analogy, that is, they should organize the controls in the same pattern as the burners (as in Fig. 8.1b). In this case, the organization of the controls carries all the information required. Without effort, even illiterate people know immediately which control goes with which burner, making the stove remarkably easy to use. The burners in Fig. 8.1a and b are functionally equivalent, but given considerations of usability, there is good reason to prefer the second.

Third, there are considerations related to experience, hence experiential norms. Because of their appearance, their beauty, their particular feel, their sense of quality, artefacts can make people put them to use. If two options are equal in all respects, we can expect the most experientially pleasing or most fashionable alternative to win. Designers have, quite some time now, recognized that it is not only the goal which is important but also the enjoyment of the route toward that goal. With its iPod, for instance, Apple has set a standard not only for what mp3 players ought to be capable of but also for how mp3 players should look like (small, flat, minimalistic design) and for how they should be operated (through smooth and intuitive finger scrolling rather than with annoying arrow buttons). Similarly, present-day desktop computers do no longer live up to the aesthetic imperative of cream coloredness. Car designers put great effort in getting the sound of the doors of their cars just right; slamming the door must be accompanied by a sense of robustness and quality. Usually the social enforcement of such aesthetic norms does not go through any legislative or standardization organization; they rather drive on the same kinds of social forces that any kind of fashion drives on.

Finally, the relative quality of an artefact may be a matter of cost-effectiveness. Indeed, we appreciate artefacts that are effective, reliable, usable, compatible, noninterfering, and so on but obviously not at any price. The benefits of these features should be in balance with the costs of realizing them. All else being equal, cost-effectiveness may be the decisive factor when choosing one artefact rather than the other.

A simple example can illustrate the point. It might be that a diamond blade makes a kitchen knife (slightly) more effective, or that a handmade newspaper has a better feel, or that a quadruply insulated house has a lower impact on the environment. Nevertheless, this is not how customers think these objects ought to be. The marginal—or even absent—increase in effectiveness and ergonomics and environmental soundness is disproportionate to the increase in costs. So the best knife is not the most effective in absolute terms, it is rather the one with the best balance between cost and effectiveness (or any of the other considerations above).

A summary of the norms I have discussed so far (viz., norms related to artefact use) is given in the left-hand column of Fig. 8.6. In the next section, I approach artefactual normativity from a different perspective: I detail the kinds of norms surfacing in artefact design.

4 Artefact Design and Norms

In this section, I will discuss artefactual norms that are not—or only marginally—to the concern of users but which designers have a particular interest in.

That such norms exist shouldn't surprise us. Users and designers typically pursue different ends. Roughly, a user's end is the promotion of an artefact's functional goal (+ those goals discussed in the previous section). Of course, designers contribute to the realization of the functional goals their customers have (viz., by producing the artefact), but usually, their primary interest lies elsewhere: a designer wishes to generate some kind of revenues or, more generally, to realize the goals specified in the business model she is working by. Hence, it seems quite natural that what makes for a good artefact differs for users and designers, respectively.

Nonetheless, one might think that this difference does not translate into artefacts; one might think that a designer, if she designs artefacts, conforms to the norms discussed in the previous section; she therewith has realized her own goal just as well. To design for customer requirements, so the thought runs, is all one can do to aim for financial profit. This is a mistake. In this section, I explain that the goodness of an artefact, from a designer's perspective, relates to the artefact's characteristics during its entire life cycle (Stark 2004; see also the vast literature on Total Quality Management, e.g., George and Weimerskirch 1998; Omachanu and Ross 1998). And I show that in each phase of these life cycles, norms occur that do not relate to quality, interference, nor compatibility.

So let us consider a life cycle of a product, from its design and manufacture, to service and disposal (as presented in Fig. 8.2). Each phase in the life cycle (and the norms it induces) will be discussed in the subsections below.

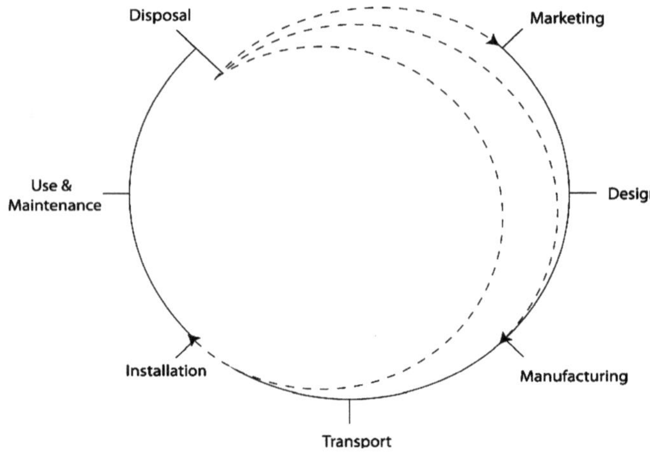

Fig. 8.2 A typical product life cycle (Adapted from Dikker 1995)

4.1 Marketability

The "marketing" stage obviously has to do with the anticipation of new customer needs and identification of existing needs. Surveying them indeed delivers essential information to design departments. It is obvious that many of the qualities we discussed in Sect. 3 enter the life cycle at this stage: designers should know what customers value and adjust their products accordingly.

But usually, a firm's marketing strategy is broader than that. Consider, for instance, a product's TTM (which stands for time to market). It is the time spent in the stages of the product life cycle in between its conception and the moment it is available for sale. Certainly for new products, TTM is important, since the sooner a company can present a first-of-a-kind product, the more it can sell before it has to share the market with competitors. The effect of TTM on a firm's income figures is presented in Fig. 8.3.

The graph shows that reducing a product's TTM usually requires extra investments. Indeed, to speed up the design and manufacturing process, more resources should be allocated to them, in the form of extra personnel, more efficient machinery, and so forth. On the other hand, these investments well may be compensated by increased revenues after the product's introduction.

Being cutting edge with respect to innovation, though, is not the only strategy to commercial success. So-called market followers may well wait for first-of-a-kind products to appear on the market before starting to design similar products, which usually differ only slightly from the original. Usually these followers compete by offering cheaper versions. For consumer products and cars, this is often easy to achieve: reverse engineers purchase competitor's products, and analyze and copy them, as such substantially reducing product development costs. In this way, the follower's strategy might be commercially attractive after all, since the product's later market arrival is made up for by lower cost figures.

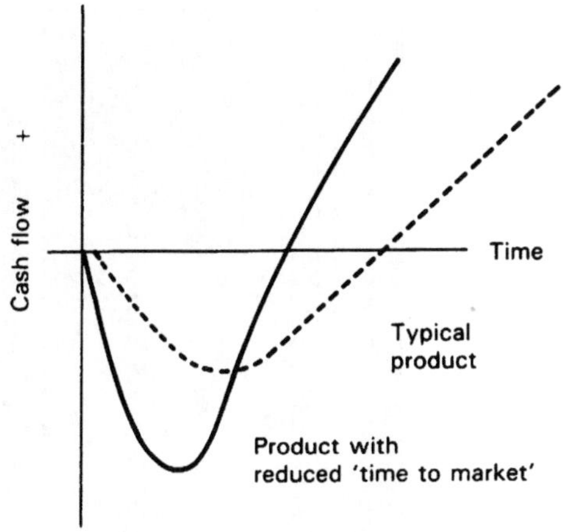

Fig. 8.3 Cost and benefits of shortening time to market

These considerations have considerable effects on design. Design decisions should be in line with the firm's market strategy. If one aims for profit through revolutionary innovation, one should design products with short TTMs; on the other hand, profit through market following allows for more modest innovations (such as lower product price) and less ambitious TTMs.

Moreover, there is a close relationship between quality and TTM. If more time and resources can be spent on the design, evaluation, and manufacturing process, more deficiencies can be corrected for and more alternatives can be tried out (Dikker 1995). Conversely, it might be a strategy—often applied in fast-moving industries—to shorten TTM at the risk of compromising quality. The point then is that engineers need to make trade-offs between quality and TTM, the latter obviously not being a design end users typically worry about.

The point generalizes for designers products ought to be marketable, that is, conform to the market strategy adopted by the firm. This implies that artefacts must meet requirements not subsumed under the class of quality, interference, nor compatibility.

4.2 *Manufacturability*

According to Gardiner and Rothwell (1986) it is crucial to consider manufacturability early in design. Manufacturing problems should be anticipated so as to avoid unnecessary and costly reiterations of the design exercise. Hence, ease of manufacture is a consideration designers should design for, irrespective of client need. Of course, there is an indirect relation between manufacturing costs and customer satisfaction, since the former are recouped from those who purchase the goods in question. Yet, typically it rewards the difference between sales price and production cost to be as high as feasible; as such, cost reduction is an aim in its own right.

Design for manufacturability means one takes into account several considerations. To begin with, one needs to keep an eye on the availability of resources, such as raw

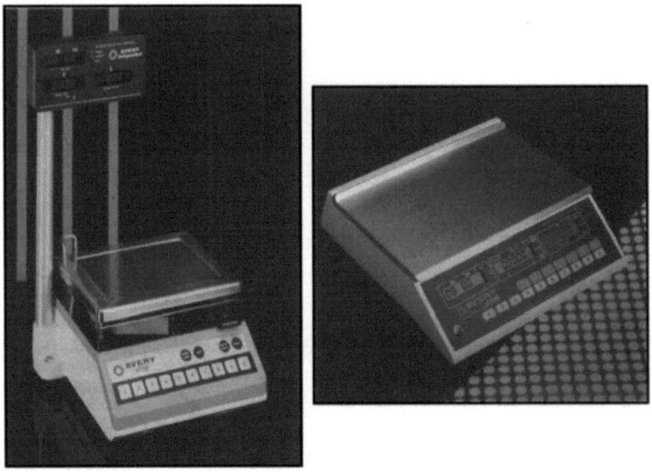

Fig. 8.4 The Avery 1770 (*left*) and the Avery A600 (*right*) redesigned for ease of manufacture

materials or components supplied by other firms. For instance, if it takes too long to obtain these materials or if the supply is unreliable, it might be better to look for alternative design solutions. Similarly, it may be rewarding to adapt one's design to the specifications of parts that are produced by known and approved suppliers or with the specifications of shared components within the firm.

The fact that engineers do not only design for function but also for manufacturability is quite clear in a case study presented by Corbett et al. (1991). It concerns the design of simple scales, used in the retail business. In 1983, W&T Avery Ltd. introduced the Avery 1770, a scale that soon could be seen in many shops throughout the UK. Although one expected world market volumes to grow, the company also expected competition to increase. Therefore, it immediately started to redesign the 1770 (see Fig. 8.4), explicitly bearing ease of manufacture in mind. The effort led in 1986 to a new scale, the A600 (see Fig. 8.4).

The Avery 1770 and the A600 are functionally equivalent, though their design differs substantially. The Avery 1770 was redesigned according to two principles. First, the configuration was made such that it allowed for a so-called layered assembly procedure, meaning that all components are added to the base in only one direction, namely, vertically down (see Fig. 8.5). Because of such assembly procedure, the machinery in the production line takes up less space, meaning that a smaller work area is needed. Second, and more importantly, the amount of components was reduced (see Fig. 8.5). This, in turn, reduced assembly time and furthermore brought along the following benefits (from Corbett et al. 1991): (1) fewer bought-in parts reduced the operational burden on the purchasing department with corresponding cost reduction; (2) bank charges were reduced as a result of fewer transactions by the purchasing department; (3) inventory was also reduced, resulting in a reduction in storage space and part transport within the plant; (4) the production control's task was much eased, as was the risk of late delivery; (5) with fewer parts, the risk of deviation was proportionately reduced, thus enhancing quality assurance; (6) with less assembly work to do, the assembly task could be conducted in a smaller work

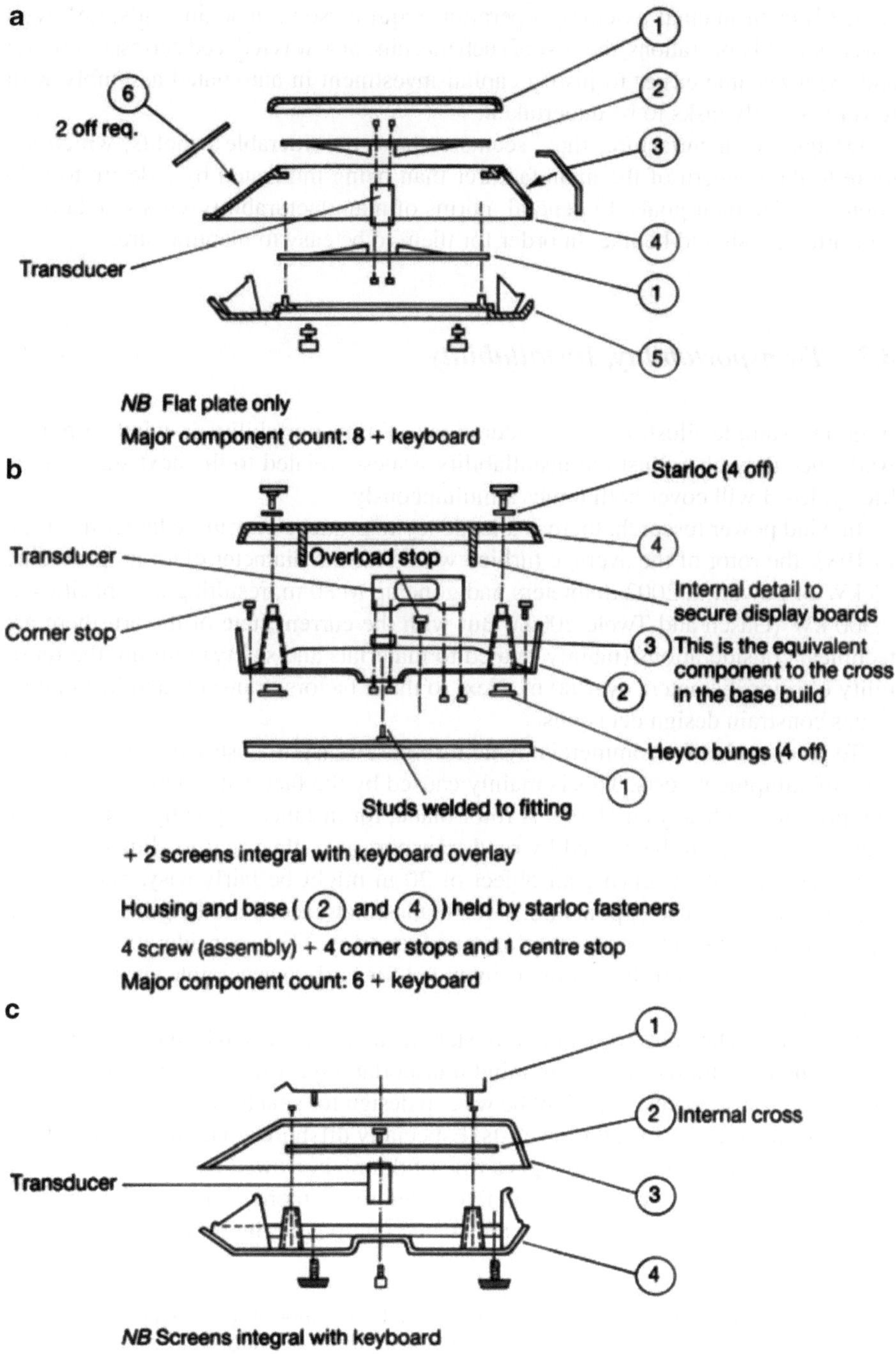

Fig. 8.5 Three options for the redesign of the Avery 1770. Option (a) requires most components, and moreover doesn't allow for layered assembly (components 3 and 6 should be added from a different angle than the other components). The winning option (Option (c)) has only 4 major components and can be assembled one directionally

area; (7) even manual assembly operation requires some machine aids, and with fewer assembly operations, the cost of such machine aids was reduced correspondingly; and (8) it became easier to justify capital investment in automated assembly with fewer assembly tasks to be undertaken.

Design for manufacture, thus, seems to have considerable benefits, which are more to the concern of the manufacturer than being motivated by a desire to help clients realize their goals. In general, norms of manufacturability set a standard for how artefacts should be like, in order for them to be easy to manufacture.

4.3 Transportability, Installability

A good example illustrating considerations of transportability is wind turbines.[4] And since they also illustrate installability issues—related to the next stage in the life cycle—I will cover both topics simultaneously.

In wind power research, there is a tendency to produce ever more larger turbines. In 1983, the rotor of the average turbine would have a diameter of 15 m, providing 55 kW of power; in 2002 diameters had gone up to 80 m, resulting in capacities of 2,500 kW (Gasch and Twele 2002). But with the current state of the art, there are technical considerations (mainly related to materials and safety) limiting the feasibility of rotor diameters over 80 m. Next to these factors, transport and installation issues constrain design decisions.

To give an idea for commercial wind turbines, transport costs can consume up to 20% of equipment costs. This is mainly caused by the fact that much of a turbine's components are heavy and large. A rotor blade, for instance, typically comes in one piece. So if it has to be moved by road transport, it is clear that its dimension will play a crucial role. Carrying an object of 20 m might be fairly easy, but one can expect that costs raise exponentially for blades that exceed the length of a regular truck trailer or that are broader than a typical roadway. These considerations become especially important if the turbine is intended for fairly inaccessible sites (Manwell et al. 2002).

So even if technical considerations as stability and safety could be met for diameters over 80 m, one still has to keep in mind transportability; if revenues do not cover the increase in transport costs, it might be wise to design for smallness.

The same goes for installation costs. Especially offshore wind turbines should be designed for ease of installation. As a rule of thumb borrowed from the offshore oil and gas industry, any work on sea is five to ten times more expensive than on land (Kühn 2002). The reasons for that are scarce equipment, high mobilization costs,

[4] Another nice example concerns Baird's discussion of the evolution of spectographs, where issues of transportability proved to be crucial (see Baird 2004). More generally, Baird argues that progress in scientific experimentation, and thus in science in general, was accelerated once instruments were developed that could be transported to the site where they would be used that the increase in transportability of instruments seriously increased the reach of science.

weather delays, and safety matters. For instance, for the installation of one wind turbine, one might need up to seven dedicated barges and vessels.

The main strategy to reduce costs is to adapt turbine design to the difficulties of offshore operations. This means that the designed turbine should have as few as possible heavy components to be lifted and that it allows for off-site precommissioning (e.g., it should be possible to assemble the rotor or tower in the harbor). Both factors reduce installation time and cost and thus increase commercial viability.

One might think that the above only applies to large equipment, but also simple consumer goods often bear witness to concerns of transportability and installability. Most IKEA products, for instance, are designed in such a way that the amount of space needed to transport them is minimized; all components of a piece of furniture should be easy to install and assembly, and should neatly fit in a box that is as flat and small as possible.

4.4 Recoverability

From the last two phases in the life cycle, we can derive criteria we already encountered in Sect. 3. First, related to "use and maintenance," we can discern usability and maintainability as common customer requirements designers design for. In this stage, we presumably do not find anything that is exclusively to the concern of the designer.

The "disposal" stage involves considerations of interference; sustainability, among other things, means that also during the passive stages of their life, artefacts should minimally interfere with other entities, such as the environment (cf. norms of interference, Sect. 3.2). In as far as consumers and/or legislation care for the environment, this is certainly something to design for. But recovering discarded goods can be regarded as a business opportunity as well, one that allows companies to generate profits. This perspective induces specific challenges for designers.

Discarded goods can be reused as sources of energy, as secondary materials, as spare parts, and so forth. To make such reuse profitable, however, one needs to design for ease of disassembly and recoverability. For that purpose, BMW, for example, set up in the late 1980s a pilot recycling plant in Landshut (Siuru 1990). In the factory, engineers examined techniques to disassemble cars and to recover parts and raw materials. And even more importantly, they registered the bottlenecks in the process, information that was used later for the design of future generations of BMW automobiles. According to Penev (1996), these efforts led to the following new norms for design: (1) the number of materials used in a car should be limited as much as possible, (2) composite elements should be avoided, (3) the number of nonreversible joints should be reduced where possible, (4) parts and materials should be marked to allow easy identification during disassembly, and (5) only recyclable plastics should be used.

Also design for non-disassembly might promote recoverability. Disassembly and dismantling of audio equipment, telecommunication products, car stereos, and the

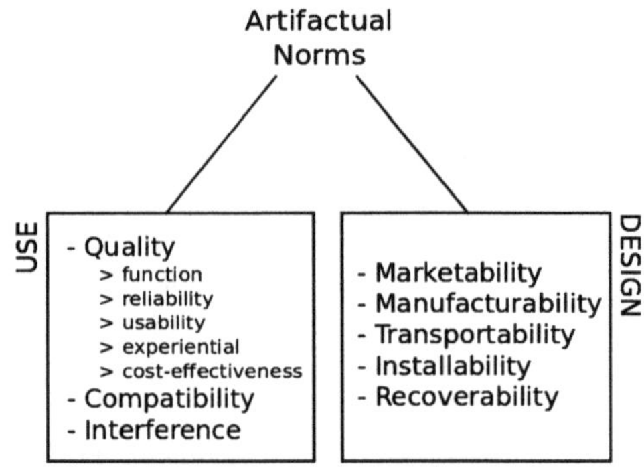

Fig. 8.6 Artefactual norms

like, is often a costly operation. However, one can design them in such a way that they, once discarded, can be processed as a whole. Ram et al. (1998), for instance, discuss design for disassembly in the consumer electronics industry. In their example, mechanical processing of electronic equipment results in material fractions which are suitable for further processing in copper smelters, aluminum smelters, incinerators, and so forth.

5 Discussion

Figure 8.6 gives an overview of the norms discussed above. Each category involves institutionalized evaluative standards, against which an artefact's performance might be set. An artefact may be good, for instance, in meeting common standards of quality or compatibility, or by conforming to prevalent norms regarding the recoverability of discarded goods, or by being state of the art in terms of manufacturability, and so forth. The goodness of an artefact, thus, is broader than its goodness at performing its proper function—pace standard treatments concerning the normativity of artefacts.

I think there is much to be gained from such a broader perspective on artefactual norms. Functional norms might be sufficient to characterize idealized contexts, in which artefacts are nothing more than simple function bearers. In contemporary settings, however, artefacts are typically not just simple tools; they have (in fact, need to have) various other qualities just to be able to compete the multitude of other artefacts on the market and to be able to safeguard the other values and goals we cherish.

Apart from that, my view sheds new light on design practice. It is common to characterize the work of designers as the exercise of determining a physiochemical structure for a prespecified function (see, e.g., Vermaas 2006; de Ridder 2006; Kroes 2006). In light of my argument, this seems too unconstrained to be an accurate picture of what engineers typically do. It is reasonable to suppose that the function of cars has remained fixed over the years; if so, what on earth have car designers been doing if its function they are mainly working on?

Here is definition of design, taken from a recent engineering textbook, that is more responsive to my concerns:

> Engineering design is the systematic, intelligent generation and evaluation of specifications for artefacts whose form and function achieve stated objectives and satisfy specified constraints. (Dym and Little 2005)

This conception has three advantages. First, function and form are put on a par, instead of form necessarily following function. Second, Dym and Little do not define the objectives of artefacts in functional terms; they recognize that artefacts may satisfy other human ends. And third, engineers design within a given set of constraints. Again, these need not be functional. In contemporary car design, for instance, a car's function is taken to be fixed, while the engineer's main task is to find reasonable trade-offs between norms related to, such things as, aesthetics, ease of manufacture, environmental soundness, and so forth. Resolving conflicts between these issues is what car designers have kept busy over the last years (decades, century).

Finally, if one aims to understand contemporary artefacts, one cannot ignore the vast amount of nonfunctional norms and standards drafted by such institutions as ISO, DIN, NEN, and so forth. The wide variety of product norms that I have discussed help to protect and guide customers, they help to constrain the design space that engineers need to consider, they can be invoked in case of legal disagreements, they help to reduce switching costs, and so forth. Their role is quite crucial to current technological practice. Any suitable philosophical account of technology, I think, should be responsive to that fact.

References

Baird, D. 2004. *Thing knowledge – A philosophy of scientific instruments*. Berkeley: University of California Press.

Blind, K. 2004. *The economics of standards: Theory, evidence, policy*. Cheltenham: Edward Elgar.

Bratman, M.E. 1987. *Intention, plans, and practical reason*. Cambridge: Cambridge University Press.

Bucciarelli, L. 1996. *Designing engineers*. Cambridge, MA: MIT Press.

Corbett, J., M. Dooner, J. Meleka, and C. Pym. 1991. *Design for manufacture: Strategies, principles and techniques*. New York: Addison-Wesley Publishers.

Dancy, J. 2004. *Ethics without principles*. Oxford: Oxford University Press.

de Ridder, J. 2006. Mechanistic artefact explanation. *Studies in History and Philosophy of Science Part A* 37(1): 81–96.

de Vries, H.J. 1999. *Standardization – A business approach to the role of national standardization organizations*. Boston: Kluwer Academic Publishers.

Dikker, F. 1995. A knowledge-based approach to evaluation of norms in engineering design. Ph.D. thesis, University of Twente.

Dym, C., and P. Little. 2005. Engineering design: A material and processing approach. McGraw-Hill Publishers.

Farrell, J., and C. Shapiro. 1988. Dynamic competition with switching costs. *RAND Journal of Economics* 19: 123–37.

Franssen, M. 2006. The normativity of artefacts. *Studies in History and Philosophy of Science Part A* 37(1): 42–57.

Gardiner, P., and R. Rothwell. 1986. Design management strategies. In *Technology strategy and the firm: Management and public policy*, ed. M. Dodgson. Harlow: Longman Publishers.

Gasch, R., and J. Twele. 2002. *Wind power plants: Fundamentals, design, construction and operation*. London: James and James.

George, S., and A. Weimerskirch. 1998. *Total quality management: Strategies and techniques proven at today's most successful companies*. New York: Wiley Publishers.

Hardcastle, V.G. 2002. On the normativity of functions. In *Functions: New essays in the philosophy of psychology and biology*, ed. A. Ariew, R. Cummins, and M. Perlman. Oxford: Oxford University Press.

Hilpinen, R. 2004. Artefact. In *Stanford encyclopedia of philosophy*, ed. E.N. Zalta. URL: http://plato.stanford.edu/entries/artefact/. Last accessed 3 Feb 2010.

Kroes, P. 2006. Coherence of structural and functional descriptions of technical artefacts. *Studies in History and Philosophy of Science Part A* 37(1): 137–151.

Kroes, P., and A. Meijers. 2006. The dual nature of technical artefacts. *Studies in History and Philosophy of Science Part A* 37(1): 1–4.

Kühn, M. 2002. Offshore wind farms. In *Wind power plants: Fundamentals, design, construction and operation*, eds. R. Gasch, and J. Twele, 365–384. London: James and James.

Manwell, J., J. McGowan, and A. Rogers. 2002. *Wind energy explained*. New York: Wiley.

Neander, K. 2004. Teleological theories of mental content. In *Stanford encyclopedia of philosophy*, ed. E.N. Zalta. URL: http://plato.stanford.edu/entries/content-teleological/. Last accessed 3 Feb 2010.

Norman, D.A. 1988. *The psychology of everyday things*. New York: Basic Books.

Omachanu, V., and J. Ross. 1998. *Principles of total quality*. Boca Raton: CRC Press.

Penev, K. (1996). Design of disassembly systems: A systematic approach. Ph.D. thesis, Eindhoven University of Technology.

Ram, A., J. Deckers, and A. Stevels. 1998. Recyclability of consumer electronics – Design for non-disassembly. In *Proceedings of Care Innovation 1998*, ed. B. Kopacek, 137–147. Vienna: Austrian Society for Systems Engineering and Automation.

Raz, J. 1975. *Practical reason and norms*. Princeton: Princeton University Press.

Simon, H. 1969. *The sciences of the artificial*. Cambridge, MA: MIT Press.

Siuru, B. 1990. From scrap heap to showroom. Mechanical Engineering, November.

Stark, J. 2004. *Product lifecycle management*. New York: Springer.

Vaesen, K. 2006. How norms in technology ought to be interpreted. *Techné* 10(1): 97–115.

Vaesen, K. 2008. A philosophical essay on artefacts and norms. Simon Stevin Series in the Philosophy of Technology. Ph.D. thesis Eindhoven University of Technology.

Vermaas, P.E. 2006. The physical connection: Engineering function ascriptions to technical artefacts and their components. *Studies in History and Philosophy of Science Part A* 37(1): 62–75.

von Weizsäcker, C.C. 1982. Staatliche Regulierung – positive und normative Theorie. *Schweizerische Zeitschrift Für Volkswirtschaft und Statistik* 2: 325–343.

von Wright, G.H. 1963a. *The varieties of goodness*. London: Routledge and Kegan Paul.

von Wright, G.H. 1963b. *Norm and action*. London: Routledge and Kegan Paul.

Walton, D.N. 1990. *Practical reasoning: Goal-driven, knowledge-based action-guiding argumentation*. Savage: Rowman & Littlefield.

Chapter 9
Instrumental Artefact Functions and Normativity

Jesse Hughes

1 Introduction

Artefacts are inherently practical things, intended to be used to achieve certain kinds of ends. This is, after all, what we mean when we speak about artefactual functions that things of this sort are good *for* something or, to put the matter differently, that things of this sort are good for *something*—that is, that the function bearer is good at achieving certain related classes of ends. To possess a function is to be suited for certain specified things.

Moreover, there is a natural relationship between artefactual functions and practical reasoning. More often than not, we are interested in knowing an artefact's function so that we will know how it can be used to achieve various ends which we have now or may later adopt. A sailor is interested in the function of his tiller because he needs to know how to use it to steer the boat. Functional knowledge thus has practical consequences: a sailor who knows how to steer using the tiller has a much greater likelihood of reaching his destination than the unfortunate soul who finds himself alone on a boat without the same functional knowledge.

In other terms, the sailor has gained a clearer understanding of the means available to reach his goal than the marooned incompetent. As a result, he is more likely to succeed, but this knowledge also comes with normative consequences: if the sailor fails to use the tiller in the proper manner, then we should fault his judgment. If our marooned incompetent, on the other hand, does not use the tiller properly (and so does not reach his destination), it is surely not his fault.[1] He does not know how to

[1] In this dire circumstance, we may well think that the marooned incompetent should try to learn how to steer on his own, if need be, but that is another matter and his failure to teach himself steering is not comparable to the sailor's failure to steer when he already knows how.

J. Hughes (✉)
Bentley University, Waltham, MA, USA
e-mail: jhughes@bentley.edu

M.J. de Vries et al. (eds.), *Norms in Technology*, Philosophy of Engineering and Technology 9, DOI 10.1007/978-94-007-5243-6_9,

steer, while the sailor does. To put this obvious point clearly, what constitutes rational behavior in each circumstance depends on what the agent knows about the circumstance, and this includes knowledge about relevant artefact functions.

Nonetheless, the prevailing theories of function are not well-suited for understanding the normative consequences of functional knowledge. Philosophical discussions of functions have largely focused on their contribution to theoretical analyses, not practical reasoning. This is because much of the philosophical interest in functions is due to their recent reintroduction in biology. Consequently, functional theories aim either to explain the presence, prevalence, or persistence of function-bearing features (especially biological), as in Wright (1973), Millikan (1989), and Neander et al. (1991), or to explain a capacity exhibited by a larger system by a component-wise analysis, as in Cummins (1975). In both cases, functions are useful for the explanations they provide, that is, for their contributions to our theoretical knowledge rather than for any more directly practical consequences.

Because we are interested in both agent and artefact norms, rather than in explanations, instrumental function (introduced in Hughes 2009a) is a more natural concept for our analysis. Instrumental functions are intended to simplify the step from functional claims to practical consequences, by focusing precisely on the relevant practical aspects of certain kinds of functional knowledge. The previously mentioned theories of functions serve a different role: they are primarily concerned with the use of functions in scientific contexts. Instrumental functions, on the other hand, emphasize functional claims in practical contexts—interactions between user and artefact, the connection between artefact performance and normative claims (such as whether the artefact is well- or malfunctioning), function creation and publication in engineering, and so on. For a more thorough comparison of Millikan- and Cummins-style functions to instrumental functions, see ibid.

2 Instrumental Functions

What are the features of (certain) function ascriptions that allow one to derive clear practical consequences? That is, how does one step from a claim, such as "Bolt cutters are for removing padlocks," to an intention to use *this* pair of bolt cutters in a particular way so that *this* particular padlock can be removed?

We take for granted an instrumentalist model of practical reasoning, in which such reasoning combines an existing desire with beliefs about (relevant) causal relations to yield an action, an intention to act, or a normative judgment regarding an action, depending on the author and context (we tend to express the conclusion as a normative judgment). To take an example from von Wright (1963)

I want to make the hut habitable.
Unless I heat the hut, it will not be habitable.
―――――――――――――――――――― (1)
Therefore, I must heat the hut.

In this example, the first premise is a claim that I desire to make a particular proposition to be true—that is, that I want to attain a particular *end*. The second proposition expresses a means-end relation, namely, that heating the hut is a necessary means to making it habitable. The conclusion here expresses that practical rationality requires that I heat the hut.

What role, then, should functional knowledge play in an instrumentalist account? How can knowledge about an artefact's function be used in a syllogism like the above? By and large, we expect that the effect must be in the second premise—instrumental functions matter to us because they provide new information about means to our ends. Artefacts are valuable to users because they can be used to attain existing ends, not because they generate new ends.[2]

Thus, the first bit of practical information conveyed by a functional claim is that the relevant artefact can be used as a means to its functional end. To continue with our previous example, if Ethel knows that bolt cutters are for removing padlocks,[3] then she may reason thus:

I want to remove this padlock
Bolt cutters are for removing padlocks.
___
Therefore, I have a reason to use bolt cutters. (2)

Our second premise is very different than the corresponding premise in von Wright's syllogism, which alleged that heating the hut was a *necessary* means to our agent's end. In this case, on the contrary, all we know that bolt cutters *may* remove a padlock. Bolt cutters are not, presumably, the only way to remove padlocks.[4]

As a consequence, the conclusion here is weaker than that of von Wright's syllogism (1). We do not conclude that Ethel *must* use the bolt cutters, since there may well be other means capable of removing the padlock. She may achieve her goal by one of these other means. Nonetheless, knowing that bolt cutters can be used thus gives Ethel a reason to use bolt cutters.

Let us make the reasoning more explicit. If one knows that

Bolt cutters are for removing padlocks, **(FN)**

[2] That's not to say that functional knowledge never results in new desires. At least some advertising aims to create consumer desires by providing functional knowledge ("It slices! It dices! It bathes the room in the soothing aroma of fresh peaches!"). We want—and are persuaded to want—artefacts because of their functions. Nonetheless, this relationship between artefact functions and desires is tangential to our investigation here, though it may well be central to an account of innovative engineering. See also the discussion of maieutic ends in engineering in Hughes (2009b).

[3] Of course, bolt cutters have other functions, such as cutting through chain link fences, but let us grant that removing padlocks is a function of bolt cutters.

[4] More precisely, bolt cutters provide a (possibly weakly) sufficient means. For a discussion of syllogisms involving sufficient means, see Audi (2006) or Railton (1997). The syllogisms we discuss below are primarily derived from the latter.

then one may conclude that

Using bolt cutters is a sufficient means to removing padlocks. (ME)

We take statement (FN) to be a(n incomplete) functional ascription and statement (ME) a means-end relation[5] which apparently follows from (FN). Thus, we may restate (2) in more explicit means-end terms:

I want to remove this padlock.
Using bolt cutters is a sufficient means for removing padlocks.

Therefore, I have a reason to use bolt cutters. (3)

This, indeed, seems a promising reconstruction of the sort of reasoning we use. I infer the means-end relation (ME) from the functional ascription (FN) and conclude that I therefore have reason to use bolt cutters.

And yet, the conclusion does not appear to be a normative judgment regarding a particular action but rather a judgment regarding a *kind* of action. We have not concluded that we should use *this* pair of bolt cutters or *that* pair but rather that we should use *a* pair. When it comes time to act, however, I must proceed with a particular pair and not an unspecified pair. We need to step from the general norm here expressed to a norm regarding a particular, clearly specified act.

There is a similar, more subtle general/specific mismatch in the premises. The first premise expresses a desire to remove a particular padlock, while the second expresses that our tool is capable of removing padlocks in general. If (ME) actually meant that the bolt cutters in question could remove *any* padlock, then clearly the reasoning goes through, since we could replace the second premise with its consequence:

Using bolt cutters is a sufficient means to removing *this* padlock. (ME′)

Alas, it is unreasonable to think that (ME) really means that *any* pair of bolt cutters is sufficient to remove *any* padlock. One should always use the right tool for the job, and there is good reason that bolt cutters come in different sizes. A well-documented bolt cutter has a specific maximum capacity it can cut through—the higher the capacity, the more expensive, larger, and heavier the pair, generally speaking. What Ethel wants to know is that

Using *this* pair of bolt cutters is a sufficient means to removing *this* padlock. (ME″)

[5] Following von Wright (1963), we adopt the convention that means are actions, such as using an artefact, and ends are states or conditions one may desire to attain.

Insofar as she can determine that it is, then she has reason to use this pair. Thus, in what circumstances does the functional ascription (FN) justify the inference to (ME″)?

It should be clear at this point that the sentence written as (FN)—what we often take to be the functional claim—is incapable of supporting the inference here. That sentence does not include the information needed for our conclusion, and so it seems that the sentence (FN) leaves implicit some of the functional knowledge necessary to draw reliable practical conclusions. In fact, in order to conclude (ME″)—or even to determine that this pair of bolt cutters is a *relevant* consideration—one needs answers to the following questions regarding the functional claim:

(a) What kind of function bearers is this about (and is this pair of bolt cutters one of them)?
(b) What result should they produce?
(c) In what situations should they work (and is this situation one of those)?

In addition, in order to actually act appropriately, we must answer one further question:

(d) *How* do I use it?

These questions correspond to the primary features of instrumental functions, respectively:

A. The artefact type
B. The functional goal
C. The specification of contexts of use
D. The use plan

We will give a brief overview of each of these in turn. For more details on these features, see Hughes (2009a).

2.1 *Artefact Type*

For our purposes, instrumental functional ascriptions specify the function of function-bearing types—especially, though not exclusively, artefact types. Even in informal functional talk, the artefact type is typically explicit. For instance, it is clear that "Bolt cutters are for removing padlocks" connects a functional goal (removing padlocks) to an artefact type (bolt cutters).[6]

Artefact types may be broad or narrow, depending on the context. We may speak of the function of bolt cutters generally, or the function of heavy-duty, steel-handled, 24″-length brand X bolt cutters. The breadth of the artefact type helps determine the specificity of the remaining instrumental function characteristics. We may say, for instance, that the brand X bolt cutters are for removing padlocks with a shackle no greater than 5/16″ diameter, a much more specific claim than we make regarding

[6] In unusual circumstances, one may find functions that apply to particular tokens rather than a broader type. We are here interested in type-level instrumental functions, however, since they are more relevant for exploring artefact normativity. See Hughes (2009a).

bolt cutters generally. The breadth of the artefact type thus acts as a *level of abstraction*, in the sense of Floridi and Sanders (2004).

There are natural restrictions on the breadth of the artefact type. Because instrumental functions include use plans, functional goals, and contexts of use, the artefact type must be narrow enough so that each token is used in the same way, to achieve the same goal and in the same circumstances. There are, for instance, many different kinds of wine bottle openers, from a traditional corkscrew to the syringe and air pump variety to cordless electric openers with push-button interfaces. Clearly, the use plans for these disparate types are very different: you do not use an electric bottle opener in the same way you use a corkscrew. Thus, the type "wine bottle opener" is too broad for our interests here, because its tokens are associated with different use plans. The artefact type must be narrow enough so that its tokens share the same fundamental function characteristics.

The type should also be broad enough so that we do not distinguish tokens of the same basic design. Two electric bottle openers of the same make and model may differ because they were manufactured at different locations or at different times. We should surely treat these as tokens of the same type, regardless of our level of abstraction. On the other hand, a 2009 (model year) Jetta may be considered a different type than a 2010 Jetta, since the two differ in more ways than just manufacture date. We may distinguish the two model years as different types or identify them as a single type (Jetta), depending on our needs and interests.

We use T to denote the artefact type of a function ascription.

2.2 Functional Goal

Instrumental functions are practical. They express the suitability of using a kind of artefact in order to bring about a certain state of affairs. In terms of means-end reasoning, the functional goal identifies certain ends which are attainable by using a token of the appropriate type in the appropriate situations and in the appropriate manner. This capacity is the source for the instrumental value of an artefact type—we value bolt cutters *because* they remove padlocks (or cut through fences, and so on). If bolt cutters had neither this nor any other instrumental function, then they would have no instrumental value as a type, although particular tokens may still be used (and hence appreciated) in ad hoc manners (as paperweights in remarkably breezy offices, say).

We use φ to denote functional goals, since the state of affairs toward which our usage aims may be expressed as a propositional function. We do not assume, however, that such goals are Boolean, that is, that one either realizes the goal or not. Some goals (such as, "Stopping the car quickly") can be satisfied to greater or lesser degree. We also do not assume that the user can reliably determine whether (or to what degree) the goal has been realized, but we *do* assume that there is a fact of the matter involved. When I take antibiotics, I have no way of knowing whether they produce the desired effect (even if I feel better, that may be a result of my natural immune system rather than the antibiotics), but either they were effective or not (or were effective to some degree), regardless of my personal ignorance.

It is worth emphasizing here that we view functional goals as *propositional functions* and not simple sentences.[7] In logical terms, they include (typed) free variables: bolt cutters are not for removing particular padlocks, but for removing padlocks (or padlocks of a particular type, say, with suitably small shanks). This functional goal may be crudely represented as *Remove(x)*, where x is a variable ranging over padlocks of the appropriate type. A particular use of the bolt cutter for this purpose will aim at removing a particular padlock s and so will be successful just in case, afterward, *Remove(s)* is true.[8]

2.3 Contexts of Use

Artefacts are designed to work in particular circumstances. A particular design of bolt cutter, for instance, is intended to cut through shanks of a particular range of diameters. Complete knowledge of an artefact's function includes not only *how* but *when* the artefact ought to be used. Restrictions on the intended *contexts of use* place limits on when one can reasonably expect to achieve his goals by using the artefact. A 24″ pair of bolt cutters is intended to cut through a 5/16″ (or less) padlock shank. When I try to cut through a 1/2″ shank and fail, I cannot complain that this pair of bolt cutters has not done what it should.

Thus, each instrumental function comes with a set C of situations in which the artefact type is intended to be used—the *specification of contexts*. A particular usage occurs in a particular context c. We write $c \in C$ to indicate that c is a situation satisfying the specification C.

Specification of contexts serves three distinct roles:

(i) They limit the situations in which an artefact is expected to perform its function. A car should not be expected to provide reliable transportation if its operator does not know how to drive (does not have *operational knowledge*, in the terminology of Houkes 2006).

(ii) They provide parameters for the use plan and functional goal. When we use a pair of bolt cutters to remove a padlock, the padlock s is part of the context of use, and the use is successful if and only if it realizes *Remove(s)*.

(iii) Success can be context dependent. Brakes should stop cars on both wet and dry pavement, but we expect shorter stopping distances on dry pavement.

---

[7] In this respect, our account is similar to Millikan's relational functions. See Millikan (2002, 1984).

[8] For a development of instrumental functions in a more formal setting, see Hughes (2005).

2.4 Use Plan

Finally, instrumental functions involve particular actions. When we say that bolt cutters are for removing padlocks, we mean that there is a particular way to use bolt cutters in order to effect that end. Instrumental functions come with *use plans* (see Houkes 2006; Vermaas and Houkes 2006). One does not know an instrumental function (in a practical sense) unless she has some idea *how* to use the artefact in order to realize its functional goal.

We denote the use plan α. Like the functional goal, a typical use plan includes typed parameters. For instance, using a pair of bolt cutters to remove a padlock involves positioning the cutters so that the shank of the lock is between the blades and pushing the handles together. If x ranges over tokens of type *Boltcutter* and y over *Padlock* (of the appropriate size), then our simple user plan may be represented as *Position(x,y); Close(x)*. A particular application of a pair b of bolt cutters to a particular padlock s would be denoted *Position(b,s); Close(s)*.

In practice, use plans may be either explicit and detailed or vague and broad. The amount of detail depends, in part, on the current level of abstraction. They may include conditional actions ("If the stapler is empty, load it."). But in each case, they describe what one should *do*. Thus, use plans provide the *means* for our means-end analysis. Roughly, then, things which are *done* are part of the use plan while *end states* toward which the action aims are part of the functional goal.

2.5 From Functions to Means

As we have argued previously (Hughes 2009a), instrumental functions are characterized by the four features described above. Thus, we will use the tuple $\langle \varphi, \alpha, C, T \rangle$ to denote the instrumental function with artefact type T, functional goal φ, context specification C, and use plan α. An instrumental function ascription in its essence, then, entails the following:

In situations c satisfying C, one can use a T-token t as prescribed by $\alpha_{c,t}$ in order to realize $\varphi_{c,t}$. **(FM-1)**

At least, this *would* be the case, if not for the fact that different T-tokens may behave differently. Bolt cutters can be sharp or dull, and sharp bolt cutters are more likely to achieve their end than dull.

For this reason, we introduced the concept of "normal tokens" in (Hughes 2009a). Given a functional ascription $\langle \varphi, \alpha, C, T \rangle$, a *normal token t* of type T is one with the physical features required to realize φ in the manner intended—that is, one which has the physical features specified by the artefact type's *design.*[9]

With this notion at hand, we amend (FM-1) as follows. Given a instrumental function ascription $\langle \varphi, \alpha, C, T \rangle$, one may conclude that

In situations *c* satisfying *C*, one can use a *normal T*-token *t* as prescribed by $\alpha_{c,t}$ in order to realize $\varphi_{c,t}$. **(FM-2)**

In the next section, we will begin to unpack some of the normative consequences of this entailment. Note, however, that (FM-2) does *not* entail the instrumental function ascription $\langle \varphi, \alpha, C, T \rangle$. The statement (FM-2) is purely about *capabilities* of (normal) *T*-tokens. Instrumental functions, on the other hand, include an *intentional* aspect. Not only must *T*-tokens be capable of realizing φ as specified above but they must be *valued* for this reason. Gasoline-fueled internal combustion engines can be used to contribute greenhouse gases to the atmosphere, but this is not what they are *for*. Loosely speaking, in order for φ to be the *function* of type *T*, someone must value *T*-tokens for their capacity to realize φ.

For a detailed discussion of the teleological nature of instrumental functions, see Hughes (2009a).

3 Prescribed Usage: Reasons for Action

We see now the most direct, practical consequences of instrumental functions. Knowing that an artefact type has an instrumental function entails a family of associated means-end relations. This, in turn, produces reasons for the agent to act appropriately, assuming that he wants to realize the functional goal in some suitable context. As with other (weakly sufficient, in the terminology Hughes et al. 2005) means, the normative conclusion of such practical reasoning is fairly weak: the agent simply has a reason to use the handy artefact token, but this reason is defeasible. It is not irrational to forgo the usage, assuming that other means exist to achieve his desired end.

[9] Of course, not every artefact type has a formal specification, but we nonetheless presume that there is a sense of normal token available for each type—more precisely, for each function. Carpenter's hammers are for pounding nails. They vary widely in their specifications, but in every case, they have a handle roughly perpendicular to a head. The head should be strong enough to strike an object without fracturing or coming loose from the handle. A hammer which fails to satisfy these basic restraints is certainly not a normal carpenter's hammer and is unlikely to serve its function of pounding nails well.

In this and other cases, one may infer what features are relevant for "normalcy," by analyzing how the type is expected to realize its function. A token with the requisite features is normal, and one without is not. As one might expect, designers, engineers, and others with deep technical knowledge will have a more precise grasp of what counts as a normal token than everyday users.

We may now give the explicit syllogism for drawing practical consequences from functional knowledge. In order to sketch the reasoning in natural language, we will use "T-tokens are for realizing φ," to express the instrumental function $\langle \varphi, \alpha, C, T \rangle$, leaving C and α implicit:

I want to realize $\varphi_{c,t}$.
T-tokens are for realizing φ.
t is a normal token of type T. **(PU)**
c is a situation satisfying the specification C.

Therefore, I have a reason to do $\alpha_{c,t}$.

Let us call a syllogism of this form a *prescribed usage* of t. We may illustrate this with our padlock example once again. In this case, I may reason thus:

I want to remove this padlock.
24″ bolt cutters are for removing small padlocks.
This artefact is a normal pair of 24″ bolt cutters. **(PU^b)**
This padlock is a small padlock.

Therefore, I have reason to use this artefact to remove this padlock.

If I believe each of the premises is true, then I cannot help but accept the conclusion.

Note, however, that I am not practically irrational if I choose *not* to use this pair of bolt cutters, although I *am* irrational if I do nothing (and intend to do nothing in the future) in order to remove the padlock, all the while still desiring its removal.

My knowledge of the function of bolt cutters thus allows me to infer that using bolt cutters is a means to certain ends. Typically, they are one means among many. Depending on my skills and other tools available, for instance, I may consider picking the lock as an alternative means to my end. I may have both 24″and 36″ bolt cutters at my disposal and be free to choose between them. A rational agent in this situation has several options. How shall he decide among them?

This decision problem is not, of course, unique to reasoning about artefacts. In most situations in which we pursue a given end, we have choices to make about the means to that end. We do not aim, therefore, to give a full account of how one selects (or ought to select) one means among many here but instead will illustrate a few of the relevant factors (reliability, effectiveness, and costs and side benefits) that play especially important roles in artefact selection.[10]

---

[10] These factors, taken together, may be used to determine a preference on means and hence determine which mean is optimal, in the sense of Audi (2006), though this would require some effort.

Reliability. Given an end φ, an action α is a more *reliable* means to that end than another action β if doing α is more likely to realize φ than doing β. Thus, reliability is a measure of the likelihood of success. In this situation, if α is otherwise equivalent to β, then the agent has a stronger reason to do α than to do β. Suppose, for instance, that either 24″ or 36″ bolt cutters may remove the padlock at hand, but the longer bolt cutters are more likely to succeed than the shorter, due to mechanical advantage. In this case, if I have both sizes of bolt cutters available and I am otherwise indifferent between the two, then I have a stronger reason to use the 36″ bolt cutters rather than the 24″, because the former are more reliable.

Effectiveness. In the same situation, if α realizes φ to a greater degree than β, we say that α is more *effective* than β. All other things being equal, this gives one a stronger reason to do α rather than β, provided the difference in degree *matters* to our agent. On my folding bicycle, for example, I can use either a hand brake (which brakes the front tire) or a backpedal brake (which slows the rear). Both brakes have the same functional goal, namely, to slow or stop the bike in a timely manner, but the hand brakes slow the bike more quickly than the backpedal brake. Sometimes, I strongly desire to stop the bike quickly. On those occasions, I use the hand brakes (or, better, both brakes simultaneously). On other occasions, a gentle stop is sufficient and so I have no stronger reason to use the hand brake over the backpedal brake.

Cost and side benefits. The bulk of practical deliberation, however, concerns other features that distinguish α from β. Actions have side effects that make them attractive or unattractive to the agent. Those effects which are unattractive we refer to as *cost*. This includes literal monetary cost, energy usage, pollution or noise produced, and so on. Other effects are valued by the agent, and we refer to these as *side benefits*. This includes simple enjoyment in performing the action, the achievement of other ends pursued by the agent, and so on. Such considerations often play the determining role in selecting an appropriate action. If I do not need to slow my bike too quickly, I prefer to use the backpedal, because the hand brake can increase the risk of losing control of the bicycle (since it grabs the front wheel of the bike, rather than the rear). Similarly, a competent sailor who needs to arrive in Martha's Vineyard from Cape Cod in a day or so may choose to sail rather than take the (more reliable and faster) ferry, because he enjoys the ride. Cost and side benefits are obviously a crucial consideration in selecting one's course of action.

This is a broad overview of the simplest practical consequences of instrumental function. Knowing the function of an artefact provides one with certain associated means-end relations. These relations are weakly sufficient, in the sense that they assert proper artefact usage *may* (in some cases, *will*) realize the functional goal. The practical normativity involved in such knowledge is fairly weak: it gives the agent a reason to use the artefact to achieve his ends in appropriate circumstances, but the strength of the reason depends on comparing the reliability, effectiveness, and cost and side benefits to other alternatives available to realize that same end.

4 Prescribed Usage: What the Artefact Should Do

We turn our attention now to norms regarding artefact tokens in a prescribed usage. Given the instrumental function $\langle \varphi, \alpha, C, T \rangle$, a token *t* is "supposed to" be capable of realizing φ in a prescribed usage. This fundamental normative claim allows one to distinguish between good (i.e., working or well-functioning) and bad (i.e., malfunctioning) *T*-tokens.

We should note here that this distinction is about *capabilities* and not *actual performance*, although the two are certainly related. A perfectly good token may, in a prescribed usage, fail to bring about φ. Some artefact types are not expected to realize their functional goals in every prescribed usage. An antiaircraft missile, for instance, may miss its target although it is working properly. Thus, the fact that *this* missile failed to hit its target is not, by itself, proof that this missile is malfunctioning. The question is whether this missile is *capable* of hitting its target—or, more precisely, whether it is as likely to hit its target as it ought to be.

This raises the question: how likely ought success to be? A properly sharpened pencil that fails to leave a mark on writing paper seems to be a bad (not to mention fairly puzzling) pencil, while the failure of an antiaircraft missile to strike its target is acceptable. Thus, and not surprisingly, our answer must depend on the artefact type.[11] A well-functioning token is as *reliable* as normal tokens of the same type in the same context of use.

Similarly, a well-functioning token is as *effective* as a normal token of the same type in the same context of use. If a car's brakes are incapable of stopping the vehicle as quickly as a "normal" set of brakes would, then they are not functioning as well as they should.

Thus, we propose the following definition of malfunction[12]:

A token *t* is *malfunctioning* with respect to a proper function if it is unable to realize φ as *reliably* or *effectively* as normal *T*-tokens in some situations *c* satisfying *C* when used according to α, that is, if $\alpha_{c,t}$ is not a reliable or effective means to $\varphi_{c,t}$. **(Mal)**

A token which is not malfunctioning is *well-functioning*.

Note that this is a fairly narrow definition of malfunction. We treat only the negative aspects of malfunction: a token malfunctions when it cannot do what it is supposed to do. There is another side of malfunction, namely, a token may malfunction when it does something it should not do. A car that emits more pollution than it should is malfunctioning, although it nonetheless fulfills its obvious function: it is

---

[11] And the function of interest, in the case of types with multiple functions.

[12] First discussed in (Hughes 2005) and presented in greater detail in (Hughes 2009a).

a reliable means of personal transportation. Let us leave aside this alternative sense of malfunction here, to be dealt with in later work.[13]

It follows from definition (Mal) that normal tokens are well-functioning. One may suppose that the converse is true as well, that "normal" and "well-functioning" are synonyms, but this is not the case. Consider, for example, a set of rabbit ear antennas for a television set. Sometimes, one of the "ears" breaks off, resulting in poor reception, but a moderately resourceful person can fix this by attaching either a wire hanger, aluminum foil, or both. The resulting contraption is surely still a pair of rabbit ears, though it is not a *normal* pair. Nonetheless it may work adequately well. If so, it is a non-normal but well-functioning token of its type.

As we said, actual performance does not, by itself, entail malfunction. The fact that a token fails to realize its functional goal in a particular usage does not mean that the token is malfunctioning.[14] Nonetheless, repeated failure is certainly a good reason to suspect malfunction. Suppose, for instance, that I hit the power switch on my television and nothing happens. It does not turn on as I expect it to. How should I react?

Most likely, before concluding that the set is broken, I would try hitting the power button a couple more times. Perhaps it did not engage like it should have (either because the power button itself is a malfunctioning component or because, sometimes, even simple buttons fail to engage as they should). Let us suppose that, again, nothing happens. The set is not functioning as it should, as far as I can see.

I look at the set. I considered it a normal token previously and it still appears normal to my (uneducated) eyes. Perhaps, then, I was mistaken about the context of use. The obvious question is whether the set is currently powered. Is it plugged into a powered wall socket? If so (and if nothing else suggests that the context of use is not appropriate), then I must conclude that this is, indeed, a prescribed usage.

At this point, when I have confirmed repeated failure in prescribed usage, I have reasonable evidence of malfunction. I could still be mistaken in this conclusion—the failures could be due to some statistical fluke,[15] for instance, or I could be wrong

---

[13] Franssen (2006) gives an account of malfunction that effectively includes both negative and positive senses: "'x is a malfunctioning K' expresses the normative fact that x has certain features f and that because of these features, a person p has a reason not to use x for K-ing." If a car emits too much pollution, then this is a reason not to use the car, and so it is malfunctioning in Franssen's sense.

[14] Note that if normal tokens are expected to always realize the functional goal, which may be the case with some particularly simple artefacts, then a single failure would indeed indicate that the token at hand is not as reliable as normal tokens and hence is malfunctioning.

[15] If a normal token realizes its goal with reliability less than one, then no finite number of failures will prove beyond doubt that the token is malfunctioning, though the longer the sequence, the greater the probability that it is due to malfunction rather than statistical happenstance.

about the actual context of use—the wall socket may not deliver the right voltage, perhaps. If, as it turns out, the set is malfunctioning, then it is not, contrary to appearances, a normal token of its type. It may well appear normal to me, but I am wrong. A trained engineer or repairman should be able to discover how the set fails to be normal.

Notice that, once I come to the conclusion that the token *t* at hand is malfunctioning, any usage of *t* is *nonprescribed*. The syllogism (PU) includes the premise that *t* is a normal *T*-token, and this premise is false if *t* is malfunctioning. Thus, let us turn our attention to nonprescribed usage.

5 Nonprescribed Uses

In this section, we will sketch some initial considerations about norms and nonprescribed uses. We are not here prepared to give a full defense of these normative claims but rather to simply give an initial starting point to the discussion of nonprescribed uses and responsibilities.

Suppose that I am aware of the instrumental function $\langle \varphi, \alpha, C, T \rangle$, but that I am considering a usage which is *not* prescribed, in the sense of Sect. 3. There are various ways in which a usage may not fit the form of syllogism (PU)[16]:

- The token *t* at hand is not a normal *T*-token.
- The context of use *c* does not satisfy the specification of contexts *C*.
- My planned use of *t* is not consistent with the use plan α.

In each case, the difference between prescribed and nonprescribed usage is a matter of degree. The non-normal token may be more or less similar to normal tokens, the context may be more or less similar to the specifications, and my planned usage may be more or less similar to the accepted use plan.

In general, one expects that the more similar the actual situation to the prescribed usage, the greater the likelihood that the nonprescribed usage will be successful, but the details matter. If I wish to cut a small pipe and my only cutting tool I have at hand is a wood saw, I may be tempted to use the wood saw. Unfortunately, this is not a very good plan. Wood saws are not well-suited to cut metal, and the saw will likely be damaged in the attempt. On the other hand, if I want to cut a small piece of wood and all I have is a hack saw, it is not such a bad idea to use the tool at hand. It should suffice for cutting wood. In order to judge whether a nonprescribed usage is reasonable, one relies on his judgment and experience to a greater extent than in the prescribed case. Instrumental functions

[16] We omit one possibility: the end which I am pursuing is not a functional goal for any artefact type *T*.

are, in some respects, like moral rules of thumb: we rely on our functional knowledge to avoid more tedious and difficult reasoning in order to realize our goals. When functional knowledge does not apply to the situation at hand, practical conclusions are harder to draw.

An agent often *does* have good, practical reasons to use an artefact in a nonprescribed manner. Ethel's bolt cutters may be designed to remove only small padlocks while she needs to remove a somewhat larger padlock. In this situation, a rational agent may well conclude that "it's worth a try." She may have other options available—she may, for instance, consider buying a larger pair of bolt cutters—but nonetheless conclude that at least attempting the nonprescribed usage is preferable, since she may always pursue the alternative course on failure.

Note, however, that the norms regarding the artefact differ considerably between prescribed and nonprescribed usage. Ethel's bolt cutters are not *supposed* to remove *this* (somewhat larger) padlock. This is not what they were designed for. Thus, failure in this case is *not* good evidence of malfunction. We should not blame a token for failure to do what it was not designed to do.

The responsibility for nonprescribed usage falls squarely on the user, then, and not the artefact (or its designers). When an artefact fails in a prescribed usage, we may regard it as the "fault" of the artefact. It ought to have worked. This is even clearer in the case of costly failure: Ethel's bolt cutters shouldn't break when used to remove a small padlock. But, if they break when used to remove a larger padlock, then we are more likely to blame Ethel, rather than the cutters. Ethel has acted unwisely by using the cutters thus.

Nonprescribed usage may violate more serious norms than mere practical self-interest. In some cases, nonprescribed usage puts property or life in danger. Airplanes are supposed to be flown by persons trained in the skill. This is part of their context of use. If I attempt to fly an airplane despite my lack of training, then I am certainly engaged in a nonprescribed usage that is not likely to realize my goal—I'm not likely to get where I'm going in one piece. Moreover, I am putting others at risk—especially if I somehow manage to get airborne.

Even experienced pilots may crash a plane, of course, but the moral harm here is more significant. An experienced pilot is using the plane as it should be used. I am using a plane in a nonprescribed manner. Because of this fact, I assume more responsibility for harm done. When we deviate from use prescriptions, we either increase the risk or lower our capacity to assess the risk, or both, and are thus a natural subject of greater moral responsibility.

Of course, these moral judgments do not alleviate the designer of all responsibility. A well-designed artefact will perform reasonably well in circumstances that are close to prescribed usage, and known dangers regarding nonprescribed usage should be communicated to the user. Notwithstanding these issues, the more abnormal the usage, the more responsibility falls on the user's shoulders.

Clearly, there is more to be said regarding norms in nonprescribed usage. We leave a more thorough examination of this issue for later work.

6 Concluding Remarks

Knowledge about artefactual functions is inherently practical. Such knowledge produces practical reasons for using the artefacts as well as norms for judging artefact tokens. These features of functional knowledge are essential to understanding how persons reason about artefact usage and artefact performance, and traditional functional theories (which focus on theoretical explanations rather than practical reasoning) do not make this connection explicit. We have shown here how the theory of instrumental functions, on the other hand, does yield clear practical consequences. Viewed in this way, functional knowledge adds to the available means to certain ends.

To be sure, one could reason about each particular token individually and come to essentially the same practical consequences. Even if Ethel doesn't know what bolt cutters are for generally, given time and cleverness, she could discover that this particular pair of bolt cutters is well-suited for removing padlocks. Of course, this sort of ad hoc reasoning is inefficient to the point of absurdity. We pass on functional information because it serves as an effective rule of thumb for discovering new means to our ends. If Ethel knows what bolt cutters are for generally and knows that this is a pair of bolt cutters, then she has some idea how they can be used and a defeasible expectation that they will suffice to achieve their goals (assuming that they appear normal).

But we often choose to use artefact tokens in situations outside of their intended use. Such nonprescribed uses can be rationally justified, although the user implicitly accepts more of the responsibility for failure in such cases. The fact that these bolt cutters are for removing small padlocks places a limit on my expectations regarding their performance. Failure to remove a small padlock is some evidence of malfunction, but failure to remove a larger padlock—something they were not designed to do—need not reflect badly on the bolt cutters.

It is in this last area, the division of responsibility for nonprescribed uses, that is most open for new research. To what degree does the user accept responsibility for catastrophic failure in such situations? In part, it must depend on whether the particular nonprescribed use was foreseeable by the designer and the catastrophic failure preventable (either by design or by education of the user). But here, too, the devil is in the details: the manufacturer of a mass-produced and widely used product has an apparently greater responsibility to predict such (mis-)use than the craftmaker or hobbyist who produces only a few. The division of responsibility for nonprescribed uses between user and designer is obviously a subtle and difficult topic which we are unable to address fully here.

References

Audi, Robert. 2006. *Practical reasoning and ethical decisions*. London: Routledge.

Cummins, Robert. 1975. Functional analysis. *Journal of Philosophy* 72(20): 741–765.

Floridi, L., and J.W. Sanders. 2004. Levellism and the method of abstraction. The final draft of this paper is available as IEG Research Report 22.11.04. See http://www.wolfson.ox.ac.uk/~floridi/pdf/latmoa.pdf

Franssen, Maarten. 2006. The normativity of artefacts. *Studies in History and Philosophy of Science* 37: 42–57.

Houkes, Wybo. 2006. Knowledge of artefact functions. *Studies in History and Philosophy of Science* 37: 102–113.

Hughes, Jesse. 2005. Means-end relations and artefactual functions: A sketch. Presented at the *Norms, Reasoning and Knowledge in Technology* workshop, Boxmeer, the Netherlands. Available at http://phiwumbda.org/~jesse/papers/index.html

Hughes, Jesse. 2009a. An artefact is to use: An introduction to instrumental functions. *Synthese* 168(1): 179–199.

Hughes, Jesse. 2009b. Practical reasoning and engineering. In *Philosophy of technology and engineering sciences*, ed. Dov M. Gabbay, Anthonie Meijers, Paul Thagard, and John Woods, 375–402. Elsevier B.V. North Holland, in Amsterdam.

Hughes, Jesse, Peter Kroes, and Sjoerd Zwart. 2005. A semantics for means-end relations. Presented at SEP 2005. Available at http://phiwumbda.org/~jesse/papers/index.html

Millikan, Ruth Garrett. 1984. *Language, thought and other biological categories*. The Cambridge, MA: MIT Press.

Millikan, Ruth Garrett. 1989. In defense of proper functions. *Philosophy of Science* 56: 288–302.

Millikan, Ruth Garrett. 2002. Biofunctions: Two paradigms. In *Functions: New essays in the philosophy of psychology and biology*, ed. André Ariew, Robert Cummins, and Mark Perlman, 113–143. Oxford: Oxford University Press.

Neander, Karen. 1991. The teleological notion of function. *Australasian Journal of Philosophy* 74: 261–301.

Railton, Peter. 1997. On the hypothetical and non-hypothetical in reasoning about belief and action. In *Ethics and practical reason*, ed. G. Cullity and B. Gaut, 53–79. Oxford: Oxford University Press.

Vermaas, Pieter E., and Wybo Houkes. 2006. Technical functions: a drawbridge between the intentional and structural natures of technical artefacts. *Studies in the History and Philosophy of Science* 37: 5–18.

von Wright, Georg Henrik. 1963. Practical inference. *The Philosophical Review* 72(2): 159–179.

Wright, L. 1973. Functions. *Philosophical Review* 82: 139–168.

Chapter 10
The Goodness and Kindhood of Artefacts

Maarten Franssen

1 Introduction

One of the peculiar features of our discourse with respect to technical artefacts is its richly evaluative and normative character. We speak routinely of good alarm clocks and poor corkscrews and of functioning mobile phones and malfunctioning TV sets. Elsewhere, I have argued that the normative character of this discourse is linked to the fact that artefacts figure in a context of human action, more in particular a context of use (Franssen 2006, 2009a). Technical artefacts owe their existence to the goal directness of much of human life; they have been designed to be used, in order for their users to achieve certain goals or achieve certain purposes. Given that we have certain goals or purposes, the qualities of artefacts give us reasons to use them or not to use them in order to achieve these goals and purposes. It is because the qualities of artefacts are reason-giving that evaluative statements are normative; they express to what extent the specific qualities of a particular artefact give someone a reason to use it, given this person's reasonable goals. To be sure, artefacts are not the only things whose properties can give us reasons to use them or reasons to act in a particular way with respect to them (treat them with care, avoid them, and what have you). We can use a nutcracker to crack open a nut or we can use a rock that happens to be at hand. We may consider both to be good for the job of cracking this nut. However, this fact makes the nutcracker a good nutcracker but not the rock a good rock. The nutcracker is good *qua* nutcracker: it was designed for being used to crack nuts. The goodness of the rock in this context is not its goodness *qua* rock, but neither is it its goodness *qua* nutcracker, since the rock is not a nutcracker. At most we can say that the rock is a good rock *for cracking nuts with*. Using it for cracking nuts does not *ipso facto* make the rock into a nutcracker.

M. Franssen (✉)
Delft University of Technology, Delft, the Netherlands
e-mail: m.p.m.franssen@tudelft.nl

M.J. de Vries et al. (eds.), *Norms in Technology*, Philosophy of Engineering and Technology 9, DOI 10.1007/978-94-007-5243-6_10,

We should therefore carefully distinguish evaluative statements of the form '*x* is a good/poor Q', where Q is some sortal, and statements of the form '*x* is a good/poor Q for F-ing with', where F is some activity term. In statements of the latter form, Q could be any sortal. The statement is equivalent to the conjunction '*x* is a Q and *x* is good for F-ing with', and typically its information content, that what people who use it are interested in, is exhausted by its second term, '*x* is good/poor for F-ing with'. In '*x* is a good/poor Q', on the other hand, Q cannot be any sortal but must indicate a kind that has been designed to be used for some purpose. My account of normativity related to functionality is restricted to technical artefacts and their value. Statements that fall out of this are, for example, statements like '*x* is a good/poor heart'. Hearts are things to which functions are ascribed that are not linked directly to human actions: they are not for being *used* to achieve purposes. To be sure, hearts can figure in human actions, since their activity can additionally be put to use. The bumping sound that a heart makes when performing its biological function can be used for diagnostic purposes, for example, and for someone with a heart condition, the particular activity of this person's heart constrains the actions he or she can undertake. But it does not match the accepted meaning of 'use' to say that one uses one's heart to stay alive. For a discussion of evaluative statements of biological entities in comparison to artefacts, see my discussion in (Franssen 2009b).

In this chapter, I investigate in some detail how evaluative judgements concerning artefacts behave when additional functionally relevant information about an artefact is given: the kind of artefact that it is, the form in which it is used, a particular brand to which it belongs, and the like. I do this on the basis of a precise claim about this behaviour, which has been published a few years ago. This investigation will occupy the major part of this chapter. In the second part, I investigate the relevance of certain properties of the user for evaluative judgements about artefacts. In relation to this, I finally address the question whether such evaluative judgements are objective, given that the kindhood of artefacts can be considered objective, or user relative. I argue that they are largely objective, but not entirely so.

2 Evaluative Artefact Judgements and Functional (Sub)Categories

With respect to evaluative judgements, Sven Ove Hansson (Hansson 2006; see also Chap. 7 by Hansson, this volume) has recently argued that evaluative statements about the members of a functional category Q answer to a certain logic with respect to functional subcategories of Q. Any category can be further divided into subcategories according to whether the members of the category have additional attributes that define the subcategories, and so can functional categories. The functional category of cars, for instance, can be divided into the subcategories of red cars, blue cars, white cars, and so forth, or in the subcategories of cars made in Germany, cars made in Japan, and so forth. A special sort of subcategory in which a functional category can be further divided is a *functional subcategory*, which is defined by

Hansson as adding further specifications of the function defining the category. Now Hansson's claim is that if Q_i is a functional subcategory of the functional category Q, then if any (sub)category member *x* is a good Q_i, *x* is a good Q, but not the other way around: if *x* is a good Q, it is not necessarily a good Q_i. For 'poor', exactly the reverse is true: if *x* is a poor Q_i, *x* is not necessarily a poor Q, but if *x* is a poor Q, it is also a poor Q_i. Or, in logical terms, the conjunction of two assertions of the form '*x* is a good Q' and '*x* is a Q_i', where Q_i is a functional specification of Q, does not entail an assertion of the form '*x* is a good Q_i', but an assertion of the form '*x* is a good Q_i' does entail an assertion of the form '*x* is a good Q'. And for 'poor', the reverse: the conjunction of two assertions of the form '*x* is a poor Q' and '*x* is a Q_i' entails an assertion of the form '*x* is a poor Q_i', but an assertion of the form '*x* is a poor Q_i' does not entail an assertion of the form '*x* is a poor Q'. We can abbreviate this in saying that, for functional subcategories with respect to their functional kinds, 'good' only necessarily transfers 'upward', whereas 'poor' only necessarily transfers 'downward'. Only the combination of the good/poor attribute and an attribute that further specifies the functional kind is supposed to answer to this logic. For each in isolation, the attributes just combine by conjunction, as they do for any two normal properties, making it the case that a blue car that is made in Germany is a car that is blue and made in Germany. An example of a combination of the attribute 'good' and a non-special attribute is that, necessarily, a good car that is blue is a good blue car. An example of a combination of a functional-specification attribute and a normal attribute is that, necessarily, a blue car that is a racing car is a blue racing car. However, a good car that is a racing car is not necessarily a good racing car, and a poor racing car is not necessarily a poor car, or so Hansson claims.

Although Hansson does not set out to give an explanation for this logic, the following argument could be thought to underlie it. In order to classify as a Q, *x* must satisfy certain conditions, and in order to classify as a Q_i, it must additionally satisfy some further conditions. Then, in order to classify as a good Q, *x* must satisfy the conditions for classifying as a Q, and it must satisfy at least a subset of these conditions to a particularly large extent. (At least, this is what we presume: in order to be a good Q, it is not the case that *x* must be a Q and additionally must have properties that have nothing to do with its Q-hood. Something must be able to qualify as a good Q merely through its Q-hood.) In order to classify as a good Q_i, it must *additionally* satisfy the conditions for classifying as a Q_i, and it must satisfy at least a subset of *these* conditions to a particularly large extent. (This is also assumed: qualifying as a good copy of a subclass of Q must go at least partly through the properties that define the subclass.) This implies as a possibility that an item may fail to adequately satisfy one of the additional conditions for being a good Q_i while still adequately satisfying all the conditions for being a good Q. Then, being a good Q is compatible with not being a good Q_i. However, an item that satisfies all the conditions for a good Q_i satisfies all the conditions for a good Q, because the latter are included in the former. Thus being a good Q_i is necessarily being a good Q. Poorness, in contrast, corresponds rather to a failure to adequately satisfy certain of the conditions that determine Q-hood, though not necessarily the conditions that must be particularly satisfied for goodness, since goodness need not be the same as the absence of poorness.

The complement of 'poor' could be termed 'acceptable' or 'fair'. So, in order *not* to classify as a poor Q, *x* must adequately satisfy certain conditions among the conditions that determine Q-hood, and in order not to classify as a poor Q_i, it must *additionally* satisfy some of the conditions determining Q_i-hood. If an item does not satisfy one of the conditions for avoiding being a poor Q, and therefore is a poor Q, it cannot satisfy all of the conditions for avoiding being a poor Q_i, since the latter include the former. A poor Q is thus necessarily a poor Q_i. But if an item fails to satisfy one of the conditions for avoiding being a poor Q_i, thus making it a poor Q_i, it may still satisfy all the conditions for avoiding being a poor Q. So a poor Q_i is not necessarily a poor Q. What all this amounts to is the supposition that each level of a functional categorization comes with its quality standards and that quality standards accumulate when one travels down the hierarchy to lower levels.

More careful consideration, however, shows that there must be something wrong here. Hansson uses the example of '*x* is a musician' and '*x* is a singer', where 'singer' is considered as a functional subcategory of 'musician'. If *x* is a good musician, it does not follow that *x* is a good singer; *x* could be a lousy singer, but *x* is a good musician because she is, say, a good pianist. And if, conversely, it is given that *x* is a poor singer, then, for the same reason, it does not follow that *x* is a poor musician, because she may be a far from poor pianist. So '*x* is a good musician' and '*x* is a singer' do not combine into '*x* is a good singer'. However, let us change the example to musical instruments instead of musical occupations. If we have '*x* is a good musical instrument', then of course we cannot derive '*x* is a good violin', because *x* may not be a violin at all; perhaps *x* is a saxophone. And surely 'saxophone' is a functional subcategory of 'musical instrument'. But what if we have to combine '*x* is a good musical instrument' and '*x* is a saxophone'? Then, it *does* follow that *x* is a good saxophone. What good other thing could *x* be if not a good saxophone, given that *x* is a good musical instrument? So the validity of Hansson's observation about the logic of functional goodness is limited.

3 Functional Roles and Artefact (Sub)Kinds

What seems to go wrong here is that two important aspects of artefacts, that of their functional roles and that of their – functionally defined – kinds, are mixed-up. These represent two different forms of function-related hierarchical categorization, which behave differently with respect to evaluative judgements. To clarify this, some distinctions will first have to be introduced. Typically, an artefact has been designed for a particular purpose, that is, to perform a particular function, which is, in the case of technical artefacts, either to be used directly to achieve a purpose or to be installed as a component in a larger entity.[1] An ordinary kitchen knife, for example,

[1] This larger entity will typically be an artefact, but not necessarily, for example, in the case of implants in the human body.

is, to put it precisely, for cutting with, since it is not the knife that cuts but the person who wields it. A knife can also be made for being installed in a ticket-vending machine for cutting receipts from a roll of paper once they are printed; such a component knife can properly be said to cut. A knife, therefore, is a member of the functional kind 'cutter', the kind consisting of all things that have been designed for cutting (with).[2] Not any cutter is a knife, however. 'Knife' refers to what I term an *artefact kind* in a precise sense. Artefact kinds in this sense are subkinds of functional kinds.[3] Every knife is a cutter, that is, an artefact designed and made for cutting (with), but not every cutter is a knife. There are many ways to cut, and using a knife is one of them. Other ways are using threads, paired knives (i.e. scissors), saws, jets of pressurized liquid, jets of burning gases, and so forth. Of course the precise delineation depends on the precise definition of 'cutting': it cannot be just 'separating a solid into smaller solids', since this includes also tearing or crumbling. But 'separating a solid into smaller solids along straight lines' seems an acceptable candidate.[4] When this definition of cutting is accepted, then all examples mentioned qualify as cutters, although they are far from interchangeable as far as the solid that is to be cut is concerned. What distinguishes knives from threads from jets is the *operational principle* through which they perform their function. This operational principle explains why different cutters are in use for different materials: different materials require different operational principles in order to succeed in cutting them up.

When a member of an artefact kind performs its function, we can also say that it performs 'its' role.[5] The concept of a role can be used, however, to fine-tune the use an artefact can be put to, even when used according to its designed or proper function. A standard knife, for instance, can be used to cut a tomato, slice a loaf of bread, or spread some butter or cheese on the slice of bread. We can say that we use the knife in the role of tomato-cutter, bread-slicer, and butter-spreader, respectively. And just as we can judge an artefact good or poor in the performance of its proper function, we can also judge it a good or poor performer in each of the several roles it can be

[2] Here, I take a stand on how *the* function of artefacts is determined: I take it that artefact functions are determined through their design. This is quite commonly accepted. It is controversial whether the function of artefacts can also be determined by the use they are put to, irrespective of their design. I will not go into these matters here.

[3] In the philosophical literature, especially the metaphysical literature on kinds, the distinction between functional kinds and artefact kinds is not made at all. The typical term there is 'artefact kind', but the term is explained as functional kind, for example, a watch or a clock is taken to be 'anything designed to tell the time'.

[4] Interesting borderline cases remain, however. A controlled way of tearing paper by first folding it sharply will result in straight-edged pieces. This is not considered cutting, perhaps because it does not involve a cutting instrument. A brief-opening knife operates according to just this principle, but this is called cutting open a letter, although brief openers are not so easily called knives, even if they are often shaped exactly as a knife, and cutting letters open is often done with a knife.

[5] In fact, I believe that the concepts of function and role are the same and that it is merely context and tradition that determine whether the one or the other term is used. But I will not defend this view here.

put to in accord with its proper function. The knife in question, for example, can be judged to perform best as a bread-slicer, less good as a tomato-cutter, and worst as a butter-spreader. These functional roles are also what is at issue in Hansson's example of the musician being both a singer and a pianist, but poor as a singer and good as a pianist. Even 'musician' should perhaps be already considered a role that a person can play, besides 'teacher', 'church warden' and 'secretary of the local antiquarian society'. And the person who is good musician could just as well be a lousy teacher and a fair secretary.

There is obviously an intimate link between the kind an artefact is and the role or roles it can be expected to be able to perform. A thing belongs to a particular artefact kind minimally on the basis of having been designed to perform a particular role or function *through some specific operational principle*.[6] To *that* role or function, there are no alternatives. A saxophone *is* an instrument designed to make music with in the manner of a saxophone, that is, through the operational principle corresponding to and defining a saxophone. A saxophone cannot be used to perform the violin role. It *can* be used to perform various saxophone roles: melodic solo instrument, orchestal instrument, or, if you want, car horn. Likewise, a (standard) knife is an instrument for cutting through pressure exerted on a sharp edge. It cannot perform the role of cutting thread or cutting jet. It *can* perform the roles of tomato-cutter, bread-slicer, and butter-spreader, as already noted, or, if you want, the role of nail cutter, electric-wire cutter, or letter-opener.

The relation is more intimate than this, however. The functional requirements defining the design of a particular artefact kind can be sharpened to match just one of the uses an artefact of that kind can be put to, or one of the roles it can be made to perform. And indeed, there are not just knives; there are kitchen knives, bread knives, butter knives, and what have you. In fact, the number of different kinds of knives, designed to perform very specific cutting roles, including associated roles (as for butter knives) is overwhelming, as a look at the web pages of some of the world's leading knife manufacturers will show. To try to slice bread with a butter knife is to go beyond this particular kind of knife's proper function. It is not a role this kind of knife is supposed to perform.

Now, we can see that Hansson's claim that 'good' does not necessarily transfer 'downward' when combined with predicates indicating functional specification applies only when the functional specification denotes the various functional roles a member of an artefact kind can perform in accord with its intended function. A good saxophone that performs the role of a melodic solo instrument may be a poor performer of this role (say, because the loudness of some of its notes is difficult to control), but it may be a good performer when performing the role of an orchestra

[6] Minimally because much more is necessary to unambiguously define a specific artefact kind. An unambiguous definition would refer to an artefact's detailed description, its blueprint, or perhaps to a blueprint *scheme* if one wishes to allow for the classification of the enormous variety of existing brands and serial types of cell phones into the artefact kind 'third-generation cell phone from the first decade of the twenty-first century'.

instrument (say, because it is well tuned and its timbre is exactly right). A generally good knife may be a good bread-slicer but a poor butter-spreader. This logic, however, does *not* apply when the functional specification denotes an artefact kind falling under a functional kind, or an artefact subkind falling under an artefact kind. A good musical instrument that *is* a saxophone is a good saxophone, there is no other possibility. A good knife that *is* a bread knife is a good bread knife.

However, Hansson seems additionally to have overlooked that his claim that 'poor' *does* necessarily transfer 'downward' when combined with predicates indicating functional specification does *not* apply when the functional specification denotes functional roles in accord with its intended function. It is satisfied by artefact subkinds with respect to their 'superkinds': a poor musical instrument that *is* a saxophone is a poor saxophone. And Hansson is correct in pointing out that poorness does not necessarily transfer 'upward': a poor butter-spreader is not necessarily a poor knife. Neither, however, is it necessarily so that a poor knife is a poor letter-opener or a poor butter-spreader.[7]

Note that the functional roles discussed here are roles that an artefact can be used for to perform in accordance with the kind it belongs to. Here, the artefact kind is meant, not the functional kind. Not any cutter is supposed to be able to play the roles of tomato-cutter, bread-slicer, and butter-spreader; only a knife is. A cutting torch, for instance, is not; even a sword or a pair of scissors is already out. An artefact, however, can be used to perform all kinds of roles, many of which are not in accord with its – artefact or functional – kind, that is, are not uses in accordance with the artefact's proper function. A knife, for example, can be used as a screwdriver or as a reflector or as a mirror. With respect to such accidental uses, Hansson's logic does not apply, as Hansson himself of course acknowledged (meaning that I use the term 'Hansson's logic' as a proper name, merely to refer to this logic). A good knife is not necessarily a good screw-driving knife, that is, a knife that performs well when used as a screwdriver, which is in accord with the non-necessary downward transference of goodness. Neither, however, is a good screw-driving knife necessarily a good knife, which is in conflict with the necessary upward transference of goodness.

Let us see whether we can understand this contrast between roles and kinds with respect to the logic of functional goodness and poorness by looking at the explanation for this logic proposed above. If some artefact is a member of the artefact kind 'knife', then it must have a minimum set of properties or capacities. If it fails to have these, it must either be a malfunctioning knife or a fraud or a fake knife, that is, not really a knife at all. If it satisfies this minimal set, then in order to qualify as a good knife, it must have some of these properties to a large extent.[8] Next, for each of the

[7] In Franssen (2006, 2009a), I have given other examples of the same phenomenon. A poor saw may be a good saw to use for sawing gypsum blocks (used for interior walls in buildings) to size. A (relatively) poor violin may be a good one to use for outdoor performances in rainy weather.

[8] In talking about properties that an entity can have to some large or small extent, I am greatly simplifying. I am referring to *quantities* like 'weight', 'sharpness', and 'balance', which can have a value firm a range of possible values. Quantities are not properties but rather second-order attributes, whose instances are properties, such as the property of 'having a weight of such-and-such value'. This is not the place to go into such niceties, however.

roles, it is to perform as a knife, the artefact should satisfy some properties related to this role, and to perform that role well, it should satisfy these properties to a large extent. The properties that make something a good knife, generally, need not be sufficient for making it good in each of the specific roles it can be chosen to perform. This is why 'goodness' does not necessarily transfer 'downward'. However, neither is it the case that for each of these roles, the criteria for good performance should include all criteria for making something a good knife, generally, which allows a knife that does not satisfy all these criteria still to satisfy the criteria for the good performance of some particular role. This is why, contrary to Hansson, 'poorness' does not necessarily transfer 'downward' either.

With respect to subkind specification rather than role specification, on the other hand, if an artefact kind effectively becomes diversified into various subkinds that each have their individual design criteria and accordingly their individual criteria for goodness and poorness, the 'superkind' to which they belong becomes an abstraction. Given that knife-producing firms produce kitchen knives, bread knives, table knives, butter knives, fish knives, not to speak of letter-opening knives, hobby knives, scalpels, and so forth, there is no longer an operable set of design criteria for a knife as such. There is at most an abstracted set of quite general functional requirements. Hence, a particular artefact can be judged good or bad only as a member of its subkind, the kind of thing it is – bread knife, butter knife, saxophone, clarinet. The judgement that an artefact is good as a member of the next 'superkind' it falls under – knife, wind instrument – becomes purely derivative, because there are no design criteria for the superkind specific enough to allow for independent quality judgements. Our judgement that something is a good knife derives entirely from its being a good bread knife, if a bread knife is what it is, as does our judgement that something is a good wind instrument derives entirely from its being a good saxophone, if a saxophone is what it is. So it is certainly true that, logically speaking, from '*x* is a good musical instrument' and '*x* is a saxophone' it follows that '*x* is a good saxophone', but conceptually, this has the order wrong: conceptually, '*x* is a good musical instrument' follows from '*x* is a good saxophone' and 'a saxophone is a musical instrument'.

4 Artefact Kinds, Definiteness, and Instrumental Value

There are clearly limits to the extent to which this derivative of the qualifications good or poor can be pushed. A knife is a cutter but also an instrument and an artefact. It may be uncontroversial that a good bread knife is a good knife (because it is a good bread knife), but already less so that a bread knife and a good cutter, and still less so that a good bread knife is a good instrument or a good artefact. With the increasing looseness of the broader category to the specific purpose that a bread knife is meant to be sued for, the qualifications 'good' or 'poor', if they are considered to apply at all, increasingly lose their interpretation as instrumental goodness or instrumental poorness. In 'this poor bread knife is a poor artefact', the first 'poor' is clearly instrumental, but the second poor is rather more generally evaluative, suggesting that the knife has been carelessly designed, poorly made, and therefore not meriting the positive attitude that corresponds to the general sense of 'good'. This is in contrast to the instrumental sense of 'good', which is grounded in a quite particular attitude towards a thing, that of using it for a purpose.[9]

This analysis concurs with my account of the normative content of evaluative judgements concerning artefacts, as given in (Franssen 2006, 2009a). The normative content of '*x* is a good knife' is that *x* has properties that not just make it the case that *x* is a knife but also that a person *p*'s wish for doing what a (standard) knife is designed for doing with recommends that *p* uses *x* for doing this.[10] However, if it is *p*'s wish to use the knife in a particular way, which is in accordance with the proper function of a knife, that is, cutting, but is nevertheless not a way that the knife was particularly designed for being used in, then this wish does not necessarily recommend that *p* uses the knife. After all, this particular knife may happen to be ill fit for this particular way of using it. On the other hand, if *p* has a wish to do the sort of thing for which a bread knife is specifically designed, that is, slicing bread, then to say that *x* is a good bread knife means that *x* has properties that not just make it the case that *x* is a bread knife but also that *p*'s desire to slice some bread recommends that *p* uses *x* for slicing the bread. However, if this makes *x* also a good knife, then it should be the case that *x* has properties that make it the case that *p*'s wish for cutting, generally conceived – if this is what a knife, generally conceived, is designed for doing (with) – recommends that *p* uses *x* to do it. Clearly, this claim becomes difficult to assess due to the vagueness of the terms it contains. It may be that *p*'s wish requires a general-purpose knife, but *x*, being a bread knife, neither is a general-purpose knife nor is a subkind of the kind of general-purpose knives. And the claim would be even more difficult to assess if we pulled the claim one more level up, that is, if *x*'s being a good bread knife makes *x* also a good instrument, explaining its normative content as stating that *x* has properties that, when *p* has a reasonable wish for doing what instruments are designed to support, recommend that *p* uses *x* to do it. Hence, insofar as the qualification 'good' is assumed to continue to apply, it can no longer represent instrumental goodness.

It might be remarked that the analysis presented up till now may seem to be incoherent, since I am both suggesting that we can speak of a good knife or a poor knife, and that, given the enormous variety of knives there are, we no longer can. These statements should all be understood in their status as examples, however. I hold that 'knife' as such is so vague that 'this is a good knife' strictly speaking cannot have a clear meaning. When we do say this and hold it to be meaningful, we have a

---

[9] See Franssen (2009a) for a more extensive treatment of the distinction between instrumental value and value in general.

[10] This technical formulation is explained in Franssen (2009a). The second part is almost equivalent to saying that the properties of the knife make it the case that if a person *p* has a reasonable wish for doing what a (standard) knife is designed for doing with, then *p* has a reason for using *x* to do this. The 'doing what a knife is designed for doing with' is definitely awkward, but it is crucial for my argument that this cannot be rendered as just 'cutting'.

particular kind of knife in mind, a subkind of the technical artefact kind 'knife', which can be identified as 'general-purpose knife' or 'ordinary kitchen knife' or 'table knife'. But even an ordinary kitchen knife and a typical table knife already differ considerably in their properties, as do the various members of these kinds among each other. It happens easily that one claims that knives are good butter-spreaders, only to be presented with a specimen of a table knife for which one admits that the claim does not hold, because 'that was not the sort of knife one had in mind'.

5 Other Specifications: Artefact Brands and Manufacture Histories

How does this fare with respect to other subcategories that can perhaps in some sense be considered functional as well? An example is brands: is a good knife that is a Zwilling J.A. Henckels knife also a good Zwilling J.A. Henckels knife? Or even manufacture history, by which I mean that an artefact is 'made in Germany' or 'made in China': is a good car that is made in Germany a good made-in-Germany car? It might be thought that these cases should answer to the particular logic specified by Hansson, as they are subject to the same hierarchy of criteria referred to above in the explanation for this logic. A knife could be good with respect to the general criteria for knives, or with respect to the total population of knives, whereas the firm Zwilling J.A. Henckels would consider it only a mediocre knife judged by the standards that this manufacturer considers adequate for its exquisite knives. Likewise, a good car may not be considered a mediocre or even a poor car made in Germany, that is, judged against the norms that German car manufacturers set themselves, and conversely a poor or mediocre German car may be judged overall to be a good car. This observation would be illusory, however. This becomes clear right away if we replace 'German' by 'Serbian'. Which judgements follow from which judgements depends entirely on the content of the judgements. Brand or manufacture history subcategories may be considered functionally relevant to some extent, that is, convey information on how well the members of the upper-level artefact kind can be expected to perform their function, but being of a particular brand or manufacture history does mean that additional criteria have to be satisfied for being good. Rather, belonging to a particular brand or having a particular manufacture history means that the criteria for goodness belonging to the upper-level artefact kind can be adopted in an expanded or restricted form, and which of the two is the case depends entirely on the brand or the manufacture history. The adoption leaves the criteria intact as criteria for the artefact kind of which the artefact is a particular brand or for which it has a particular history of manufacture.

So we carefully have to distinguish three cases. On the one hand, there is the case of artefact kinds and artefact subkinds. A bread knife is one of the many subkinds in which the kind 'knife' has been subdivided by design criteria. There are criteria for what makes a particular knife a good bread knife, but the existence of such criteria for all the various subkinds leaves the enveloping kind without clear criteria for goodness, making it unclear whether a good bread knife is a good knife or a poor bread knife a poor knife. Nevertheless, *if it is given* that *x* is a good knife and that *x* is a bread knife, then, necessarily, *x* is a good bread knife. On the other hand, there are two closely related cases that are in contrast with the previous case. Firstly, there is the case of artefact kinds the members of which can be used in various roles supported by the kind, such as a general-purpose knife that is used for slicing bread, cutting tomatoes, and spreading butter or cheese. Here, we have no necessary connections between the goodness or poorness of a particular knife according to the criteria for a general-purpose knife and quality of the performance in a role consistent with the artefact's knifehood. A good knife can be a poor butter-spreader; a good butter-spreader can be a poor knife. Secondly, we have the case of artefact kinds coming, to stick to these examples, in brands or coming from 'design and manufacture traditions'. The criteria for goodness on the brand level or design-and-manufacture-tradition level are a subset or superset of the criteria that determine goodness for the artefact kind but, as in the previous case, do not affect these criteria. Again, there are no necessary relations between the various judgements: a good car can be a poor car 'made in Germany' (i.e. according to German manufacturing criteria), a poor car can be a good car 'made in Serbia' (i.e. good according to Serbian manufacturing criteria), a poor car 'made in Germany' can be a good car (if the German criteria are severe enough), and a good car 'made in Serbia' can be a poor car (if the Serbian criteria are lax enough). None of these three cases satisfy the logic pointed out by Hansson.

6 Goodness, Usefulness, and Use Plans

The kindhood of artefacts is of course crucial to settle their instrumental value. If we speak of a good knife, we mean that the thing is good *qua* knife, or 'as a knife'. The judgement that something is 'good as a knife' could, however, be thought ambiguous. We may of many things say that they are 'good as a knife', for example, of a glass splinter or of the removed lid of a tin can. What we mean in these cases is that these things can be used as one uses a knife, in other words, that they are useful for cutting with. *Prima facie*, the normative contents of the judgement that something is a good knife and the judgement that something is good 'as a knife' or, as it can also be put, 'makes a good knife' are identical. On my account, the statement '*x is* a good knife' means that *x* has properties that make it the case that *x* is a knife and that make it the case that a person *p*'s wish for doing what a (standard) knife is designed for doing with recommends that *p* uses *x* for doing this, whereas the statement '*x makes* a good knife' means just that *x* has properties that make it the case that a person *p*'s wish for doing what a (standard) knife is designed for doing with recommends that *p* uses *x* for doing this. The reason-expressing parts of these two explications are the same.

However, his prima facie analysis misses important aspects. Using an artefact is almost never a matter of grabbing it and let the thing do its job. For a start, you must be able to grab it: you must have hands, or an equivalent prosthesis. Then, you must be able to hold it, which is difficult if the artefact becomes red-hot while it is performing its job. And this is only the beginning of it. Therefore, for any artefact, to make it possible to use it in the first place, there must exist – explicitly or implicitly – a description which specifies what using it consists of, that is, how you have to use it and by implication what this requires of the user, and a specification what the circumstances are in which its use makes sense, that is, will lead to the intended result. These circumstances are typically environmental: they could specify a temperature range or the required absence or presence of chemical substances, of energy sources, and the like. Barring an, often miraculous, amount of luck, the use of an artefact will lead to the desired result only if the user is acquainted with these things, such that he or she can see to it that all requirements are fulfilled, and if the requirements actually are fulfilled. We can refer to this description as artefact's use plan.[11] Next to a prospective user's aims or desires, and their reasonableness, this use plan must be brought in when considering to what extent the facts about a particular artefact are reason-giving. They can be sufficiently reason-giving only for someone who has the corresponding knowledge, has the corresponding abilities and is in the corresponding circumstances. The account for explicating the normative content of evaluative judgements about artefacts can be extended to do this in the following: '*x* is a good knife' more precisely expresses the normative fact that *x* has certain properties that make it the case that *x* is a knife and that make it the case that for any person *p*, if (1) *p*'s abilities, (2) *p*'s knowledge, and (3) the circumstances in which *p* operates are in accord with the use plan for *x*, then *p*'s wish for doing what a knife is designed for doing with recommends that *p* uses *x* for doing this.[12]

Now a use plan exists only for designed artefacts. For objects that happen to be useful in the way a designed artefact is, it is up to the user to recognize this. The proposed explication of the normative content can, therefore, also be deemed defective: only someone who would know how to handle a glass splinter to cut with,

[11] I take the term from (Houkes et al. 2002) as I have at other occasions. However, the notion of the use plan *of an artefact* does not match the idea of a use plan as it was introduced there. Houkes et al. understand a use plan to be a plan that is drawn up – an activity that is to them the primary notion of engineering design – to advise someone how a particular end or goal can be realized. Such a plan may (but does not necessarily) involve the use of one or several artefacts (which may then have to be produced to make the plan executable), and for each phase where an artefact is to be used, a fragment of the larger use plan will tell the agent(s) how to execute this phase of the plan by telling them how to use the artefact. The use plan of an artefact is therefore only a part of what Houkes et al. conceive to be a use plan. To the question what the 'use' in their conception of use plan is use *of* I can see no answer.

[12] This could be thought overly restrictive. It could be argued that the features of the knife are reason-giving for someone who has the right abilities and is in the right circumstance but just does not know this. Such a person would have a reason to use the knife but would not know that he had a reason. Whether this is considered the case depends on the particular underlying theory of normativity and reason one holds. Such details are beyond this chapter. For more thoughts on this issue, see Franssen (2009a).

or is capable of doing so for having a firm enough grasp, or a tough enough skin, or hands in the first place, has a reason to use a glass splinter to serve his or her wish to cut something. But there is no general way we can add a condition to the reason-giving explication that covers all objects that could, one way or another, be used to cut with. So on closer scrutiny, the two judgements, that something is a good knife and that something is good as a knife, do differ in their normative content. For conceptual clarity, it is best to keep the judgements well apart, speaking on the one hand of things that are good or poor knives and, on the other hand, of things that are (very) useful or hardly useful for cutting (with). The latter should not be equated to 'being completely useless for cutting (with)'. A sugar cube is not useful for cutting with, but one could, in dire circumstances, succeed in cutting a rope or a thread with a sugar cube, and say that the cube made a knife, a very poor one but nevertheless a knife. One could not have succeeded at all using a tuna sandwich.

7 The Objectivity of Evaluative Artefact Judgements

The relevance of user knowledge and abilities discussed in the previous section introduces an issue that I wish finally to address, that of the objectivity of evaluative judgements about artefacts. Until now, I have spoken of the goodness or poorness of artefacts as if this is a factual matter. But it might be considered that goodness of one or other artefact is always goodness *for someone*, with particular properties and being in particular circumstances. It was the position of Georg Henrik von Wright, one of the few philosophers to write extensively about the entire spectrum of value judgements, including instrumental value, that whether a particular artefact is a good or poor specimen of its kind is an objective fact about it (von Wright 1963). According to von Wright, it is an objective fact, given the properties a particular knife has, and given the laws of nature, what the result of cutting something will be, given the properties this something has, and it is a conceptual matter, given what it means to desire some state of affairs or aim for some state of affairs, whether a user who has the goal of seeing this something cut up considers the result to adequately realize this goal.[13] Now, the recognition that apart from the properties of the artefact, user abilities and circumstances are also relevant for determining the outcome of cutting some object, say, a tomato, with a particular knife poses a problem for this view. For the result of using an artefact to remain causally determinate, a reference to the use plan for the artefact must be included: it is an objective fact, given the properties a particular knife has, and given the laws of nature, what the result of cutting something will be, given the properties this something has, provided the user

[13] Again, we simplify things. Until now, I have treated 'good' and 'poor' attributes that are either present or absent. As von Wright emphasizes, such evaluative judgements are more often relative: the functional criteria can be satisfied to some degree, such that one knife is better than another, and a particular knife may be the best of all available knives. The point judgement 'good' could then be introduced through setting a bottom line, as 'good enough'.

of the knife handles the knife in the way specified by the knife's use plan and does so in circumstances that are in accord with this use plan. However, it is now no longer a conceptual matter whether a user who has the goal of seeing this something cut up considers the result to adequately realize this goal. The user can no longer be expected to judge just the causal result of the cutting but also what this cutting required of the user and his or her circumstances. A knife that weighs 20 k but cuts perfectly when wielded with the required force (by an exceptionally strong man, or a robot arm) will nevertheless not be considered a good knife, plausibly not even by the person who managed to cut perfectly with it. For this verdict, it would make no difference if it were clearly stated in the knife's use plan that the knife requires its user to dispose of this amount of force. One cannot trivialize the relevance to the judgement of knife's quality of such outrageous user requirements by classifying it as belonging to the kind of 'knives weighing 20 kilos' or 'knives for muscular users'.

The point is not that artefacts, or rather their use plans, should never require exceptional force or uncommon abilities of their users. Many tools, such as chain saws, presuppose considerable muscle power in their users. Cars and airplanes cannot be used successfully without extensive prior training. This does not come in the way of our judging particular chains saws, cars, and airplanes as 'good'. Additionally, the futility of introducing artefact kinds such as 'knives for muscular users' does not hold good for artefacts that require uncommon abilities because they are designed for specific minority groups with deviating abilities. Instruments with push buttons that have Braille inscriptions instead of printed ones would be all but useless for the overwhelming majority of seeing people, which is, fortunately, the overwhelming majority of all people. A pair of left-handed scissors is nearly impossible to use for a right-handed person, as is a pair of right-handed scissors for a left-handed person. A pair of left-handed or right-handed scissors is a good one if its properties give a person from the corresponding group who has something to cut a reason to use it (suppressing the niceties). For artefact that do not standardly come in such ability-specific subkinds, quality judgements of artefacts assume a background formed by the abilities, including abilities for being trained, of the average, typical, reasonably adult human user. For members of ability-specific subkinds, this background of course still contains all abilities not covered by the subkind. Additionally, an artefact should not require of its user more than is necessary, given the current state of technological knowledge. And since technological knowledge increases with time, an artefact that was considered good at one moment will be assessed as poor at a later moment.

8 Conclusions

Summarizing, in this chapter, I have first investigated the way in which evaluative judgements concerning members of artefact kinds (like knives, musical instruments, and so forth) and members of subkinds of these kinds entail or fail to entail each other, and similarly for evaluative judgements concerning members of artefact kinds and judgements concerning these members as used in specific roles or as belonging to different brands or having different design and manufacture histories. I have shown that the logic governing these entailments is different for both cases, but that in neither case, it confirms to a proposal published a few years ago by Hansson. Secondly, I have argued that evaluative judgements can be understood as largely objective but nevertheless not entirely so, as they are always made against a background of average human capabilities and the state of technological knowledge.

References

Franssen, M. 2006. The normativity of artefacts. *Studies in History and Philosophy of Science* 37: 42–57.

Franssen, M. 2009a. Artefacts and normativity. In *Philosophy of technology and engineering sciences, Handbook of the philosophy of science*, vol. 9, ed. A. Meijers, 923–952. Burlington/Oxford/Amsterdam: North-Holland.

Franssen, M. 2009b. The inherent normativity of functions in biology and technology. In *Functions in biological and artificial worlds: Comparative philosophical perspectives*, ed. U. Krohs and P. Kroes, 103–125. Cambridge, MA: MIT Press.

Hansson, S. 2006. Category-specified value statements. *Synthese* 148: 425–432.

Houkes, W., P.E. Vermaas, C.H. Dorst, and M.J. de Vries. 2002. Design and use as plans: An action-theoretic account. *Design Studies* 23: 303–320.

von Wright, G.H. 1963. *The varieties of goodness*. London: Routledge & Kegan Paul/The Humanities Press.

Part III
Normativity and Technological Risks

Chapter 11
The Non-reductivity of Normativity in Risks

Niklas Möller

1 Introduction

In contrast to the rich and often streamlined literature on the process of risk analysis, the concepts grounding the practice are under-theorised. When explicitly treated, theorists' analyses diverge. On the one hand, a dominant view in risk analysis is to treat risk and safety as relatively straightforward objective natural concepts, determined by physical facts and thus fitted for scientific study. On the other hand, theorists from the social sciences argue that risk is a 'social construct', a subjective social feature rather than a feature of the objective reality.[1]

The aim of this chapter is to argue for a thesis that is often obscured or even mixed up with the polarised debate around social constructivism: the irreducible normativity of risks that cuts across the objectivist/subjectivist issue. In particular, it affects also theories that reject constructivism[2]: even if we hold that risks are real properties of the world, there are aspects of the concepts for which we have no plausible natural reduction. By comparing the problems of incorporating the normative aspects of risk and safety with similar problems for a more general class of concepts discussed in moral philosophy – thick ethical concepts –I argue that we have good reasons to accept a general irreducibility of risk and safety.

[1] Douglas and Wildavsky (1982), Wynne (1992), Slovic (2000). See Hansson (2010) for a recent analysis of the objectivist/subjectivist risk debate.

[2] That is, reject strong versions of constructivist theses that entail a denial of real risks. Many weaker constructivist theses are perfectly compatible with the reality of risk (Möller 2011). I agree that risks are such real features of the world – although I will not argue for it in this chapter and my arguments will not depend on them being so.

N. Möller (✉)
Royal Institute of Technology, Stockholm, Sweden
e-mail: nmoller@kth.se

M.J. de Vries et al. (eds.), *Norms in Technology*, Philosophy of Engineering and Technology 9, DOI 10.1007/978-94-007-5243-6_11,

Within risk research as well as applied risk analysis, the most commonly used concept of risk is a probabilistic measure, where risk is defined as the expected value of the harmful outcome.[3] Furthermore, the notion of safety is typically taken as the antonym of risk.[4] This conceptualisation of risk and safety is often perceived as a naturalistic reduction, in the sense that the concepts of risk and safety are reduced to probability and the severity of potential harmful outcome, which are considered natural, nonevaluative phenomena possible to study with scientific methods. Risk analysis, on such an account, is a purely scientific discipline. Consequently, it is often held that the analysis of the experts is complete and the task of the decision-makers is simply to follow the presented assessment.[5] Similarly, reasoning in risk communication frequently rests on the premise that people are ignorant of the real risk and that rather than indicating a deficiency in the scientists' assessments, the major objective is to adjust the perceptions of laypeople in order to narrow the gap.[6]

In this chapter, I will argue against naturalistic reductions of the concepts of risk and safety. A consequence of my argument is that the confidence in the probabilistic concept of risk and a corresponding 'add information and stir' attitude towards disagreements in matters of risk are highly problematic. Since there are irreducible normative aspects of the risk and safety concepts, disagreements involving these aspects cannot be solved solely with the methods of (natural) science. As a consequence, risk analysis must always contain an openness for the eventuality that there are normative aspects present in a risk situation that are not covered by the descriptive operationalisations and models used in risk assessment. Naturalistic reductions of risk and safety are simplifications of the complex risk reality, and the process of risk analysis should mirror this fact.

---

[3] The expectation value approach is the standard measure in probabilistic risk analysis (Hansson 2010). Some theorists go as far as claiming that this is '[t]he only meaningful way to evaluative the riskiness of a technology' (Cohen 2003, 909).

[4] This assumption may of course be questioned (Möller et al. 2006). In this chapter, I will grant the antonym usage for *comparative* risk/safety terms, and treat 'safer', 'safest', etc., as synonymous to 'less risky', 'least risky', etc. Note, however, that this does not include a position as to when something is *safe*, that is, how the predicate 'safe' should be interpreted (but see *ibid* and Möller 2011 for a discussion). Compare also with the discussion of acceptable levels of risk and safety in the main text below.

[5] This is especially evident in cost-benefit analysis involving risk. Cf. Ackerman and Heinzerling (2002) for a critical analysis.

[6] Cf., for example, Kraus et al. (1992/2000). Similarly, Leiss (2004) argues for more emphasis on risk communication and 'science translation' (p.403) into language more easily accessible to the public, reasoning that (p.401) 'more science will not solve the essential problem, which is public distrust of the risk assessments. We knew enough about dioxin risk by about 1985, for example, to make an educated guess that exposure to dioxins is not, and is highly unlikely to become, a significant risk factor [...]. Yet the science goes on, and the controversy persists, because insufficient attention has been paid to the need for conducting a fair and prolonged risk dialogue with the public about dioxins'. Cf. Hansson (2005) and Möller (2009) for critical discussions.

This chapter proceeds as follows. In the next section, after some stage-setting discussion of natural and normative concepts and contextualisation of the concept of risk, I introduce three guiding cases that, I claim, indicate that there is something lacking in the probabilistic notion of risk and safety. In Sect. 3, I discuss when we should – and when we should not – rely on such intuitive indications, arguing that the availability of an overarching theoretical commitment must be in place for us to reasonably overrule our linguistic intuitions. Thereafter, in Sect. 4, I return to the three example cases, arguing that they are instances of normative aspects of risk and safety. In Sect. 5, I argue that the project of attempting to include these aspects on a natural construal faces severe difficulties, difficulties that indicate that they are ad hoc and unable to handle future deviations. The cause of these problems, I argue, is that risk and safety are essentially action-guiding concepts. I put the principled problem of reducing action-guiding concepts to natural ones in a larger context by arguing that risk and safety belong to the class of thick concepts that has been much discussed in contemporary philosophy, and for which the reductive project has been profoundly criticised. Finally, in Sect. 6, I sketch what I believe to be, and not to be, the consequences of the account for risk analysis.

2 Concepts and Reductions

A natural concept is traditionally conceived as a concept that is invoked in scientific explanations (Little 1994; Vallentyre 1998). More precisely, a natural concept picks out a property that is the subject matter of natural science or psychology, a property that can be known by means of empirical observation and induction.[7] Physical concepts such as length and heat, chemical concepts such as gold and water, and biological concepts such as tiger and whale are all paradigmatic natural concepts on this understanding.

Natural concepts are typically contrasted with normative concepts, concepts used to express normative claims. Good, right and fair are paradigmatic examples of such concepts.[8] Contrary to natural concepts, normative concepts are action guiding. While the fact that the cup contains water is, by itself, no reason for drinking it, that an action is good (pro tanto) entails that you have a reason to perform it.

The main question in this chapter is whether risk and safety can be reduced to natural concepts alone, in the sense of being (properly) described using only natural concepts. Against typical risk objectivists, I will answer in the negative. Hence, I claim, risk and safety are concepts that do not fall entirely on the natural side of the divide. But, as I will further show, this does not entail that risk is a subjective concept.

[7] This characterisation follows Moore (1903): §25–27. Cf. also Miller (2003), 4.

[8] Note that I use 'normative' to refer both to deontic concepts such as *right*, *ought* and *permitted* and to value concepts such as *good*, *bad* and *better*, since it is the distinction between the cluster of these broadly normative/evaluative concepts on the one hand, and the natural concepts on the other, that is of interest in this chapter.

Risk and safety are paramount concepts for societal decision-making. Admittedly, what levels of risk and safety we should accept in society are generally considered political and/or moral questions not answered by the natural sciences. But even if this is acknowledged, it does not establish risk and safety as nonscientific notions any more than our legislating a minimal temperature in our office buildings makes heat a nonscientific notion. On the contrary, risk questions are often regarded as possible to settle, at least in principle, through empirical science. Scientific disciplines provide the necessary input for risk decisions, informing us when we are at risk. Traditionally, the distinction between scientific questions and normative ones is made between risk assessment and risk management.[9] Risk assessment is considered as the scientific stage where the estimations of the risks at hand are made. Risk management uses the output of the risk assessment as input for making a decision about the risk, ultimately whether to accept or reject it.

From this description, it may seem as if there is nothing strange about naturalistic reductions of risk and safety. There are, however, important differences between paradigmatic natural science concepts such as water and heat, and the concepts of risk and safety. Or, so I will argue.

The most common conceptualisation of risk and safety used in risk analysis is the probabilistic concept of risk and safety mentioned in the introduction, in which the expectation value of harm is the quantitative measure of risk.[10] This conception will form the starting point of our investigation. For a potentially harmful event, the expectation value is defined as the product of the severity of the harm and the provbability that it will occur. This is in accordance with many basic intuitions of risk. For example, if the probability of a harmful outcome is low and the outcome itself is not severe, then the risk involved is low; if the probability is high and the outcome is severe, then the risk is high.

The probabilistic concept of risk and safety understand risk and safety in terms of severity of harm and probability, two concepts that are typically given a naturalist interpretation, often in terms of deaths for harm and objective frequencies for probability. However, even if a naturalist interpretation of harm and probability are granted, there are severe problems with understanding risk as natural.[11] In order to show this, we will use three example cases as the starting point of our analysis:

(A) New versus old design. A new bridge is to be built, and there is a choice between two construction methods for its design. One method has been used for many decades with only minor adjustments, and the other uses a new, revolutionary design that seems very promising. For the first method, there are extensive

[9] Cf. National Research Council (1983) and European Commission (2003) for standard classifications.

[10] It is worth mentioning that the probabilistic notion of risk corresponds to a quantitative interpretation of risk. As noted in, for example, Hansson (2004) and Möller et al. (2006), the term 'risk' may, depending on the context, refer to such different entities as the *probability* of a harmful event, the *harmful event* itself, the *cause* of the harmful event or even the *fact* that a decision is made under conditions of known probabilities.

[11] Later, in this chapter, we will question also this naturalist assumption for harm and probability.

frequency data for all types of load situations, weather conditions, etc., whereas for the second, although heavily modelled and tested as far as possible, there is (for obvious reasons) less frequency data. The scientists agree that on their best analysis, the expected value of harm is significantly lower for the new design, but they admit that the knowledge at hand is more uncertain.

(B) Unequal distribution. A revolutionary method of building encapsulated nuclear power plants has made the population safe from even the unlikely event of a meltdown and thus decreased the total expected value of harm. The only drawback is that some internal maintenance must be made by service staff and this is very risky, having a high expected value of harm (several magnitudes above current levels for any staff).

(C) Accident or not? Let us say that the two main travel methods between the two cities are by airplane or by car. Extensive frequency data tells us that the expected value of harm for travelling by airplane is lower than travelling by car. However, many accidents on the road are one-car accidents, and we have reasons to believe that a significant portion of these are intentional (i.e. acts of suicide). For air traffic, however, pace potential terrorist acts, there are virtually no accidents that may reasonably be labelled as suicide. If we were to exclude the suicide cases from the frequency data, the expected value of harm for car travel would be lower.

In all of these three cases, it is plausible to doubt that what has the lowest expectation value of harm is actually safest.[12] In case (A), we face a version of the quite common case of difference in the quality of our knowledge: on the one hand, we have an old, tested bridge design where the knowledge of its safety properties is quite certain, and on the other hand, we have a new design about which our knowledge is less certain. We believe that, in the end, use of the new design will result in a smaller amount of total harm, but we are not certain. In this case, the question of whether the technology with the lower expected value of harm is the safest bridge design is arguably open. The nuclear power debate is perhaps the paradigmatic case, and the contemporary debate about the precautionary principle highlights the problem of how to evaluate 'uncertain information' for questions of risk and safety (Cf., for example, Sandin et al. 2002).

Case (B) focuses on the problem of distribution: is safety a matter of receiving as low an expected value of harm as possible in a population, or does the distribution also count?[13] Maybe we should not hold that a situation is safer than another if there are some people carrying significantly higher risks than others? In most societal activities carrying risks, there is an uneven distribution of potential harm. Persons

---

[12] Note that for this to be plausible, we do not need to claim that the actual risk of X is *greater* or even equal to Y when the risk according to the probabilistic notion claims that it is less for X than for Y. The weaker claim that the risks of X and Y are incommensurable – that in cases such as these we cannot say that one is lower than the other – is sufficient.

[13] In radiation protection, for example, there is an analogous, interesting tension between the two principles of *optimisation* and *dose limitation* – cf. Wikman-Svahn et al. (2006).

living close to risky artefacts such as energy production plants or heavy chemical plants take a higher risk load than persons living further away. Likewise, for road traffic safety, there is a debate about using cables for protecting the vehicle from driving off the road. The cables are successful in avoiding many potentially lethal car accidents; however, they may also be very dangerous for motorcyclists; much more so, in general, than if the motorcyclists merely went off the road. Since there are many more cars than motorcycles on the roads, the expectation value of harm is smaller when using the wires than not, yet for the motorcyclists the risk is much higher. Is it then reasonable to claim that the method is safe as long as the total expected value is kept low, regardless of the particular risks for motorcyclists?[14]

In case (C), we face the question of how to delimit the risk and safety concepts: what events should count as risk events and, hence, be included in frequency data and other means of evaluating the risk and safety at hand? If we were to take the risk question at face value, that is, include all harmful events, air travel would be safest. But it seems wrong to include also suicides in the statistics of road traffic safety. It is one thing to allow mistakes from the driver – and surely a main part of accidents derive from the human factor – but in judging what means of transportation is the safest, it seems irrelevant that it is possible to use the car also as a tool for committing suicide. It is evident that only some harms count, not others, and in the thought experiment, we may reasonably question which means of transportation is the safest.

Cases (A)–(C) indicate that there is something lacking in the probabilistic account of risk and safety. But how are we to treat these intuitions? Could it not be argued that we should jettison them rather than the probabilistic notion of risk and safety? In the next section, we will look into the role of intuitions versus our theoretical commitments in regard to our concepts.

3 Conceptual Intuitions and Theoretical Commitments

Cases (A)–(C) are examples of how we question a notion – here the probabilistic notion of risk and safety – by means of our received intuitions. Our conceptual intuitions question that the probabilistic notion in fact captures all there is to risk and safety. An initial reaction to this could be to deny the relevance of the approach itself. Although semantic clarity is important, the objection goes, it does not really matter which conceptualisation of risk and safety we use in our scientific endeavours as long as we are fully clear and consistent in our usage. The probabilistic concept of risk – as well as other possible naturalistic conceptualisations – is on this view a technical operationalisation, and as long as the decision-makers are aware of this, it does not matter that there are other ways of understanding the concept. Taken in

[14] Interestingly, neighbouring countries such as Sweden and Norway, normally taking the same approach to matters of safety, have taken diametric positions regarding road cables. In Norway, the use is forbidden, whereas in Sweden, the use of road cables is not only allowed but expanding.

isolation, such an objection could of course be both relevant and true. That the king in chess has different properties than both fairy tale and real versions of kings is not much of an objection to the chess concept in question. In the context of concern here, however, the objection is problematic, since there is a presumption, explicit or implicit, that risk analysis is focused on capturing the risk involved not in any internal-to-science-only way, but in the sense that we use it in society when we, as laymen or decision-makers, ask how the risks are distributed, which of the alternatives are the safest, etc. (Indeed, without such an assumption, it is hard to understand the above claim that it is the laymen perception of risk that should be remedied rather than the experts': if the experts and the laymen were simply talking about different things, all that would be needed is translation, not substantial belief change.) If risk analysis is trying to answer any other question, its relevance could indeed be questioned. In other words, what is important is not only that we are clear about how our terms should be interpreted but that we are using the right interpretation. Risk analysis is principally an empirical field of study, but if we are dealing with inadequate conceptualisations, our analyses may be inadequate, too.[15]

Still, one may object, in some areas of discourse where scientific definitions diverge from ordinary language use, we say that science is ultimately the judge of the correct usage. We say that peanuts are not really nuts and that Pluto is not really a planet, although most people use 'peanut' and 'Pluto' in order to make claims to that effect (e.g. 'I love nuts, especially peanuts and cashew nuts', 'Pluto is the most remote known planet of the solar system, usually ninth in order from the sun, discovered in 1930 by Clyde Tombaugh').[16] Why would not the intuitions in play be similar to such false beliefs about nuts and planets?

The key to realising the difference between these cases and the risk and safety case lies in correctly understanding the role of theoretical commitments in these claims. Without certain theoretical commitments in biology or astronomy, we will not be in a position to correctly claim that peanuts are not nuts or that Pluto is not a planet. This role of theoretical commitments is evident in Hilary Putnam's famous Twin Earth thought experiment, which he used to argue that it is our causal connection to the terms we use that determines their correct reference. In short, we are to imagine that there is a planet, Twin Earth, similar to our Earth in all relevant respects except that the water in their lakes and seas is constituted of an alien chemical compound, XYZ, instead of H2O. Twin-water is exactly similar to Earth-water in

[15] Cf. Hare's (1952) defence against the claim that we may define moral terms as we please (p. 92): [S]ince what we are trying to do is to give an account of the word 'good' as it is used – not as it might be used if its meaning and usage were changed – [the reference to usage] is final. It is therefore no answer [...] to claim that a 'naturalist' might if he pleased define 'good' in terms of some characteristics of his choice. Such an arbitrary definition is quite out of place here; the logician is, it is true, at liberty to define his own technical terms as he pleases, provided that he makes it clear how he is going to use them. But 'good' in this context is not a technical term used for talking about what the logician is talking about; it itself is what he is talking about; it is the object of his study, not the instrument. He is studying the function of the word 'good' in language; and so long as he wishes to study this, he must continue to allow the word the function which it has in language [...].

[16] The latter an explicit claim of my computer dictionary (Apple's Dictionary 2.1.2).

its functional and phenomenal qualities, but still, Putnam insists, our term 'water' refers only to H2O, not XYZ, since it is stuff consisting of the former and not the latter that we have been in causal contact with in introducing and continuing to use the term. But what makes this conclusion justified? After all, the water I have in my bathtub and the water someone else is taking a swim in across the ocean are not numerically identical. Our claim is not of numerical identity, however; rather, we are claiming that they are of the same kind, namely, the water-kind. And we are claiming this, Putnam argues, because the essence of water is its chemical structure – it is the molecular constituency that explains its functional and phenomenal qualities. In other words, it is this theoretical commitment that justifies our treating instances of XYZ as of a different substance than our water, although it plays the exact same role (linguistic and otherwise) in the lives of the Twin Earthians.[17]

In moral and political theory, the interplay between theoretical commitments and particular intuitions is often characterised by reference to the process of reflective equilibrium.[18] When our theoretical commitments and our moral intuitions in particular cases reach a stable point of equilibrium, we have arguably found the correct reference of the moral concept in question. This is supposedly analogous to the way in which our empirical data and our theoretical commitments interact in natural science as well. When our theoretical commitments regarding a scientific concept succeed in explaining (or explaining away) our empirical data, we have found, at least approximately, the shape of the scientific term. In some cases, such as with ether and phlogiston, our revised theories and further empirical data establish that the concept did in fact not refer. In other cases, we merely adjust our commitments in relation to the concept in question: fish does not include whales, the atom turns out not to be the smallest particle, and heat is identical to mean kinetic energy.

The remaining question is thus whether we have some theoretical commitment sufficient to overtrump our intuitions in the risk and safety case. Since risk and safety analysis often deals with natural science questions, for example, questions of durability, chemical exposure or epidemiological facts, there may seem to be a case to argue for a similar set of well-motivated theoretical commitments also in risk assessment that overtrumps specific intuitions. This impression is further increased by risk perception studies. Such studies have shown, for example, that people often understand the risk of exposure to chemicals and radiation as something binary – either you are exposed or not – and underestimate the dose factor (Kraus et al. 1992/2000, 309–310). Assume, for example, that someone believes that the only thing that matters is that they have been exposed to chemical toxin X, not for how long, and thus believes that their risk of harm is equally high in situation S1 which has a short exposure to the toxin as in S2 which has a long exposure. Now, if we

[17] The role of theoretical commitments for the causal theory and other contemporary theories of meaning and reference is sometimes not appreciated, but it is acknowledged by its central proponents. In extending Kripke's causal theory from names to natural kinds, Putnam himself explicitly mentions the need for theoretical commitments for the reference of both names and natural kinds (Putnam 1975, 225).

[18] The term 'reflective equilibrium' was made famous by Rawls (1971).

assume that S1 and S2 differ in no other risk relevant feature, we may indeed rightfully claim that our theoretical commitments falsify this intuition, since our consolidated knowledge of toxicology informs us that dose is highly relevant for short- and long-term damage sustained to the human body. Hence, some risk and safety questions are settled regardless of potentially contrary intuitions – namely, those that are isolated to scientific questions for which we already have theoretical commitments in place in the form of consolidated scientific knowledge.

In a similar vein, we may say that if the only risk relevant difference between two situations is that in the first, the expected value of harm is higher than in the other, we may with the highest confidence claim that the probabilistic concept fully determines the risk question. The reason is that if we grant that no other factor is relevant for the risk, we have solid theoretical commitments in place due to the statistically consolidated correlation between higher expectation value of harm and higher actual harm. We would probably claim that the reason for a person to doubt that the higher expectation value of harm determined the risk in this case is because she has not understood the concept of expectation value, that is, similar to the case of a person with insufficient chemical understanding doubting that water is H2O.

Unfortunately for the proponent of naturalistic reductions of risk and safety, however, the background rationale of these examples far from establishes the sought conclusion. The question of whether or not there are misunderstandings among laypeople about the properties of substances and systems that are relevant for risk and safety is not under debate here. The sources of doubt of the probabilistic notion of risk to fully determine the actual risk referred to in this chapter are due neither to a misunderstanding of the concept of expectation value nor to any particular scientific facts about dangerous substances. Instead, they express a doubt, given these conceptual tools and scientific facts, that the expectation value truly captures all risk relevant differences, that is, questioning the very premise that was, for the sake of argument, granted in the previous paragraph. The reductionist needs a set of theoretical commitments overruling also such cases. In the following section, the problem of putting forward such theoretical commitments will become evident.

4 The Hard Problem for the Reductionist

The hard problem for the reductionist is that whereas there are convincing theoretical commitments in place for the scientific concepts that may solve many local problems of risk and safety (of the kind exemplified in the previous section), other sources of doubt exceed the scope of these scientific commitments. A new look at the three cases reveals distinct evaluative aspects for which the case for overruling theoretical commitments is bleak indeed.

Case (A), the new versus old bridge design, illustrates the role of epistemic uncertainty for risk and safety ascription (Cf. Möller et al. 2006). Presuming that we know that a coin is non-biased, I ascribe the probability of 'heads' in the next throw the value of 0.5. Tossing a coin from a store for magicians and cheaters, I still ascribe the probability of 'heads' in the next throw the value of 0.5, since it might be biased one way as well as the other. Our certainty about the ascription is, however, very different, since in this latter case, the correct probability may be, say, 0.05 or 0.95 rather than 0.5. Risk and safety relevant knowledge may be similar either to the first or the second scenario, as our knowledge of the total behaviour of a system or substance may be very different. Hence, our best ascriptions of the probability (as well as severity of harm) may be the same, whereas the underlying uncertainty may vary. In cases of large epistemic uncertainty, there is reason to doubt that the lowest expected value also implies the lowest risk. If I am concerned about safety, I might reasonably choose to cross the bridge that has a slightly higher ascribed expected value of harm than another bridge, if the former has a lower epistemic uncertainty, since the true value may be much higher for the latter due to the larger epistemic uncertainty. Hence, my doubt reflects an evaluative aspect of risk and safety over and above the probabilistic conception.

Case (B), the unequal distribution of risks, illustrates a second evaluative aspect of the concepts of risk and safety, which we may call distributive normativity. Doubting that a system or substance that has a lower expected value of harm is safer on the grounds that it has too unequal a distribution of harm is acknowledging that distributive aspects are part of the very concept of risk and safety. Case (B) illustrates the 'many person' distributive normativity of risk and safety. This is an important aspect in societal decision-making in cases of risk and safety, since they most often are 'many person' cases. In addition, there is another type of distributive problem of risk and safety, due to the possibility of different profiles of the harm function. The risk of one system, for example, may be due to very small probabilities for a great harm (e.g. nuclear power plant), whereas in another case, it may be due to larger probabilities for a less great harm (e.g. fossil fuel power plant). The probabilistic notion may treat these in the same way, since the actual distribution matters not beyond its contribution to the expected value, but it is arguably doubtful that this is correct in all cases.

Case (C), suicide versus unintentional accidents, highlights the evaluative aspects of the concepts of risk and safety, which could be labelled border or delimiting normativity, that is, normativity due to which events should count as safety events, relevant for assessing the question of the risk and safety in the case at hand. Case (C) indicates that not all potential harms are relevant for the risk and safety: what is qualitatively identical on the level of physical harm is not necessarily identical for evaluating the risk and safety because one event is a suicide and the other an unintentional accident. Hence, deciding which events should be included in our safety assessment is an evaluative aspect over and above the probabilistic conceptualisation.

The intentional state of the agent is arguably a major contributor in deciding what to include or exclude in judging the risk. 'Seeking harm' in a literal sense, however, is not the only potential disqualifier: looser senses of the expression may also do. There are several ways in which people may use equipment and substances in unconventional and unintended ways without intentionally harming themselves: off-piste skiing outside permitted limits, removing 'restraining' safety equipment at work or impressing on others with daring displays of courage by walking on the bridge railing. In one sense, these activities and actions are all hazardous and as such potential safety issues. Therefore, those responsible for a skiing resort must consider how to handle inadmissible off-pisters, factory safety managers must deal with deviations in safety regulations, and infrastructure planners must restrict access to dangerous areas. But it is unclear and arguably dependent on the circumstances and the exact nature of the situation at hand whether the type of events should count as safety issues.

Furthermore, the evaluative 'border issues' are not restricted only to intentional states. Other aspects of the way in which the harm is caused may be relevant, although too extraordinary events are perhaps not always relevant. If there is a meteor (or even an atom bomb) hitting the highway, for example, this is arguably irrelevant for road safety as such. However, if falling meteors were to become a more likely possibility, it would indeed be relevant. Earthquake issues are not part of traffic safety in many regions in the world because they never happen (of any significant magnitude), whereas they certainly are relevant in the Los Angeles and Tokyo areas. Magnitude of harm is also a delimiting aspect: in some situations, mere bruises or slight burn injuries are not considered harms in the relevant sense, whereas in other, even less grave risk issues, they are. Taken together, border issues represent an evaluative, contextual dimension of the concepts of risk and safety that defy the probabilistic conception.

In summary, the different normative aspects of risk and safety exemplified by the three cases – epistemic uncertainty, distributive normativity and border normativity – are areas in which we do not have theoretical commitments in place sufficient for 'overruling' doubt against the naturalistic reduction. That is, contrary to what the theorists mentioned in the introduction seem to believe, our body of scientific knowledge does not include any commitments relieving us from doubt when it comes to these normative aspects of risk and safety.

5 Risk Reductionism and Thick Concepts

The reductionist may of course claim that the failure of the last section is only due to our having focused on a too unsophisticated probabilistic conception. In principle, the reductionist could object, we should be able to include modifications to the expectation value conception, such as a reasonable measure of epistemic uncertainty. Hence, the problems put forward are not a general problem for the reductionist but merely for the particular probabilistic conception.[19]

[19] For a recent example of a more complex strategy, cf. Aven and Kristensen (2005). Also Slovic (2000) includes several complex findings of how people interpret *risk* (even if the mission in these articles is different from supplying a unified notion of risk and safety).

However, it is not sufficient for the reductionist merely to point out that a natural reduction is conceivable, that it cannot be ruled out from the outset. As discussed above, more is needed: the reductionist has to argue for a set of theoretical commitments enabling us to reduce the normative aspects of the previous section to natural ones. The evident reductionist strategy is to incorporate these aspects, on a naturalist interpretation, into the probabilistic account. While this project may seem plausible at the outset, I will argue that the evident current problem for such an attempt indicates a more serious principled problem for the reductionist. I will put this principled problem in a larger context by suggesting that risk and safety belong to the class of thick concepts that has been much discussed in contemporary philosophy, and for which the reductive project has been profoundly criticised.

In order to incorporate epistemic uncertainty, distributive aspects and border aspects in a reductionist account, we have to supply a naturalist, nonnormative understanding of these aspects. The current problem for this project is that no such account is forthcoming: there is no reasonable suggestion that can play the role of a theoretical commitment that may help us choose between conflicting intuitions about these aspects and their relation to risk and safety. Without such an account, we lack a good reason for claiming that the incorporation of these aspects to the probabilistic account constitutes a naturalist reduction rather than (irreducible) normative aspects.

Epistemic uncertainty is clearly a problematic concept to pin down. There is, among theorists, no complete and accepted understanding of the concept on which to rely. There are several models, surely, ranging from second-order probability to fuzzy set theory, but they all face problems and limitations, and there is no one accepted suggestion.[20] Hence, there is no deeply held theoretical commitment in place, and we have not reached a naturalistic reduction.[21]

The problem is arguably even harder for a reductive account including also the aspect of distributive normativity. What we need is a natural account that tells us how to evaluate the influence of distributive differences for risk and safety. But here normal scientific knowledge fails to supply the solution, since even if natural science may tell us about the probability of harmful events, it has nothing to say about distributive normativity. That is, while we may describe a difference for the population in the distribution of the expected harm, we have no naturalist account of the inference from this information to ascriptions of risk and safety.

Still, the reductionist may insist that these current problems are consistent with a reductionist approach. While there is admittedly disagreement about how to describe epistemic uncertainty, this disagreement, the reductionist may argue, is an internal scientific disagreement similar to questions of, say, how a species is to be delimited

[20] Cf. Möller et al. (2006), 421–424, for an overview of common suggestions in the literature.

[21] Note that a naturalistic reduction in the sense of interest here depends on the availability of a descriptive notion that does not use evaluative terms. For example, a second-order probability, interpreted as a subjective probability that the objective probability is correct, would count as a naturalistic reduction granted there is an available descriptive measure of the subjective probability (say, by some procedure such as De Finetti's game) in addition to the objective probability.

in detail, or what exactly constitutes a planet rather than an asteroid or a dwarf planet. Similarly, we may develop different dose limits for handling differences in distributions and their contribution to the overall risk.[22]

While it must be granted that the current problem of presenting a convincing naturalist account in itself presents no deductive argument against a naturalist account, it does provide prima facie reasons for resisting a naturalist reduction. More importantly, however, it points us to a principled problem for the reductionist. This problem is perhaps most easily visible in the case of border normativity.

In the case of border normativity, the most direct naturalist strategy would be to include the discussed exceptions into the conception. For traffic safety, for example, only the harmful events that are not acts of suicide, not the result of extraordinary events such as falling meteors, and not of too small a magnitude (such as bruises), etc., should be included. That is, on this account, we should follow a two-step process where the harmful events are first singled out according to what salient types they represent, and then the risk is a consequence of the (suitably modified) probabilistic conception. The main problem with this strategy is that it represents only an ad hoc, backward-looking modification of the naturalist conception of risk and safety.[23] The criteria for the singling out of permitted events are merely the result of previously identified problems with the general probabilistic notion. But without capturing the rationale behind the modification, it fails to motivate the generality of the account. Hence, it would merely be a piecemeal, ad hoc modification, lacking the force to capture new circumstances, or distinguish between intuitions that should be taken at face value, and those we should, after all, abandon. In other words, such a naturalist conception would lack the necessary background theoretical commitment needed for us to believe in the completeness of the account.

The general problem here is that while we may add natural descriptions to our reductive account post-factum, we have no general justification in view, and hence no way of interpreting deviations from the cases we have (arguably) managed to capture with these additions. For example, that a suicide event typically is not relevant for road safety does not mean that it is irrelevant if it includes other road users. Indeed, even within standard cases not involving collateral damage, there may be an important dividing line depending on both the cause of the suicidal action and the means through which it was performed. For example, if the intention to commit suicide was caused by the fact that the driver was strongly intoxicated, and an alcohol safety interlock device easily could have prevented the possibility of the event, it is arguably a traffic safety issue.[24]

[22] I thank a reviewer for pointing to this option.

[23] A related problem is how to express these criteria using only natural concepts. For example, even if we grant that *meteor* is a natural concept, it does not follow that we can spell out, say, 'extraordinary events such as falling meteors' in a fully natural way.

[24] Similarly, even granted (for the sake of argument) a naturalistic way of describing both epistemic uncertainty and distributive normativity able to handle the cases in the previous section, it remains unclear how complex cases are to be treated – cases involving, for example, many different expectation values, epistemic uncertainty and harm distribution, or involving all three of these aspects.

On a normative interpretation of risk and safety, resisting naturalistic reductions, the reason for the above problems of finding a general and forward-looking reduction of the concept is clear: risk and safety are essentially action-guiding concepts. That something constitutes a risk is typically a reason against (allowing, using, performing) it, and the larger the risk, the stronger is that reason. Risk has what is sometimes called a negative evaluative 'direction'. Safety, on the other hand, is a sought-after property, and that something is safe is generally an honorary state of affairs. The contribution of aspects such as epistemic uncertainty, distributive differences and border/delimitation issues comes from their complex action-guiding role, the role they play in what we have, in matters of risk and safety, reasons to do. The mistake of the reductionist is to believe that there is a substantive naturalist description that may fully capture this action-guiding role.

This critique can be put in context through the debate in moral philosophy concerning so-called thick (ethical) concepts, which have been vividly discussed since Bernard Williams' critique in his seminal Ethics and the Limits of Philosophy.[25] The idea is that there are concepts that are somewhat peculiar in that they both describe the world and are action guiding. As such, thick concepts differ from water and length, which have no such evaluative quality, as well as paradigmatic evaluative concepts such as good and right, which are said to be solely or primarily evaluative or action guiding. On the traditional analysis, a thick concept thus fills a double function: it describes a feature (the descriptive, world-guided function), and it evaluates it (the normative, world-guiding function).

Examples of traditionally discussed thick concepts are cruel, brave and selfish. They are action guiding: that an action is cruel or selfish is a reason against performing it; that it is brave is a reason for it. But they also describe the action and are thus more specific than good or right: cruelty has some relation to being indifferent – or even enjoying – causing suffering, bravery has something to do with acting in the face of danger, and selfishness has something to do with looking out for oneself.

I believe that we may fruitfully include risk and safety in the class of thick concepts. A fundamental premise behind the reductionist idea is that risk and safety have specific content.[26] That severity of harm plays a central role;that a harm is not certain but may obtain; that a risk is greater the more likely it is that the harm may occur and the greater the harm, etc. – these are all more or less specific characterisations of the concept. At the same time, as discussed above, that something is safe is a positive feature of the entity, and that something carries a risk is a negative feature of it. This action-guiding character of the notion is present in all of the risk aspects discussed above. A significant epistemic uncertainty involves the possibility

[25] Williams (1985). His account, in turn, refers to a number of earlier papers by John McDowell (1978, 1979, 1981).

[26] While we have been interested in *natural* concepts (and reductions) rather than in the descriptive concepts for which the thick concept debate is typically framed, the former is a genuine subclass of the latter: every natural concept is also a descriptive concept (although not the other way around). Hence, scepticism about descriptive reductions of thick concepts entails scepticism about natural reductions.

of a larger potential harm than ascribed and is a reason against performing that action. Likewise, when there are persons risking substantial harm, it may be a negative aspect of the situation even when the total expectation value is (comparably) low. And in the case of border normativity, the reason for, say, excluding suicides in traffic safety ascriptions is the action-guiding aspect, since when judging the risk of our means of transportation, we want to avoid taking potential harmful outcomes that are irrelevant to our purposes into consideration. Moreover, and even more fundamentally, on closer scrutiny, the very central notion of harm seems to be evaluative. Generally, while an event characterised as a harm has a descriptive component, it certainly also contains an evaluation as something negative. Even more pressing in the current context, in ascribing different severities of harm, it seems far-fetched to claim that the evaluation involved in ascribing severities of harm is only science internal. Rather, the basis for evaluation is the badness of different harmful alternatives. We typically judge a broken arm as less harmful than a severe head trauma because of the importance of the avoidance of the latter rather than the former – an action-guiding rather than scientific value. And even if we assume that death is the primary harm to take into account, it is unclear if the severity of harm is the same in all such cases. Is, in all circumstances, the possibility of a 95-year-old person dying from a medical procedure of the same severity as if the patient were 25 years old?[27] It seems hard to avoid interpreting these aspects as reflecting the evaluative, action-guiding nature of the harm concept.

Originally, theorists believed that thick concepts could be analysed as a conjunction of a descriptive part and an evaluative part, by which is meant that the descriptive content and the evaluative content can be independently given.[28] 'X is courageous' could therefore be analysed as something along the lines of 'X intended to act in the face of danger to promote a valued end' and 'this is (prima facie) good-making'.[29] If this were correct, a descriptive reduction of a thick concept would, at least in principle, be possible. While something may still be said to be lacking in the understanding of the concept of courage in a person who grasped the descriptive part but took a neutral evaluative stance towards it (did not believe that courage was either good or bad), this person would be fully able to identify courage. For the concepts of risk and safety, this would mean that there indeed was

[27] This is naturally an area of severe controversy, but we should note that there are ambitious systems of measurement designed to take account of both the quantity and the quality of life generated by healthcare interventions, such as QALY (quality-adjusted life year), reflecting the practical relevance of these concerns.

[28] Stevenson (1944), 206–207, Hare (1952), 2. Also the early Blackburn seems to have favoured such an analysis in Blackburn (1984), 148–149. Note, however, that also most influential early twentieth-century cognitivist moral philosophers assumed that moral judgments could be expressed using only a few thin evaluative concepts such as good and right and that some sort of conjunctive manoeuvre for thick concepts is possible. Elstein and Hurka (2009) point this out, naming Sidgwick, Moore, Ross, Broad and Ewing – an impressive list indeed – as companions in guilt. Another interpretation has been to treat a thick concept as a purely descriptive concept (e.g. Ayers 1936, 110–114; Mackie 1977, 41).

[29] Catherine Elgin calls this the 'skeleton account' in her (2005), 343.

a descriptive part that one could – in principle – isolate and operationalise. Perhaps an agent who understood only the natural part of risk and safety would not have a full understanding of the concept, but in terms of getting the extension of the concept right, she would not be lacking. And that is exactly what the reductionist aspire in the case of risk and safety.

In the last decades, however, this assumption has been met with an increasing scepticism. Philosophers such as John McDowell, Bernard Williams and Jonathan Dancy have argued that no such separation of a descriptive, autonomous part and an evaluation can be given.[30] The only way to understand a thick concept is to understand the descriptive and evaluative aspects as a whole. The idea is that for a thick concept, the evaluative aspect is profoundly involved in the practice of using it; one cannot understand a thick concept without understanding also its evaluative point.[31] Therefore, they have claimed, against the traditional account, that no set of descriptive terms can completely fill in the 'along the lines' of, say, 'X intended to act in the face of danger to promote a valued end' – it is disjunctively open-ended. Consequently, today most if not all theorist in the debate have abandoned such a reductive view.[32]

Applied to the context in this chapter, the output of recent moral philosophy is scepticism against the very reductive claim that lies behind the probabilistic conception of risk and safety. In effect, they argue that for this kind of concept, one has to grasp the evaluative – or, in my preferred terminology, the normative – points of the practice in which it is involved. For risk and safety, thus, there is an essential interdependence between the natural aspects and the normative aspects. That an event has a higher expected value of death than another event – treated as a natural input – is typically a reason to ascribe the former a higher risk than the latter. But this is not the case if there are other harms than death that should count as well, or a questionable distribution of the harmful outcomes for the latter case (as in the staff in the novel nuclear plant), or if the epistemic security is much higher in the latter case, etc. In other words, the three types of normativity in cases (A)–(C) are illustrations of normative concerns that are part of the very essence of the concepts, not something that can be 'peeled away' to reach a nonnormative core. From the broader outlook of the recent development in ethics, this failure of natural reductions of risk and safety is a corollary of the general failure of descriptive reductions of thick concepts.

[30] McDowell (1978, 1979, 1981); Williams 1985; Dancy (1995, 2004). See also McNaughton and Rawling (2000) and Little (2000).

[31] The rationale for calling an action *cruel* rather than merely describing it in more neutral terms is to tune in to this evaluative aspect. Cf. McDowell (1981).

[32] Richard Hare defended his reductionist strategy as late as in Hare (1997), for example, p. 61. Simon Blackburn, however, who holds a similar view in his 1984, later abandoned that view and explicitly denied it in his 1998.

6 Consequences for Risk Analysis

In this chapter, I have argued against natural reductions of risk and safety. In this final section, I will make some concluding notes about what I believe to be, and not to be, the consequences of my argument. Let us start with the latter. First, many theorists in the risk debate seem to believe that if a concept is not a natural concept, it is subjective rather than objective. It is important to note, however, that this fails to appreciate the large debate in philosophy concerning the status of the field of the normative, in particular the moral domain. In fact, many of the main proponents of the irreducible normativity of thick concepts mentioned above, for example, McDowell and Dancy, are moral realists who believe that there are objective truths in moral discourse. Hence, claiming that risk is normative is not the same thing as claiming that it is nonobjective.

Secondly, it would be a mistake to believe that my argument entails that there is no place for trying to find natural criteria corresponding to the concepts of risk and safety. On the contrary, natural criteria are both useful and important, as long as they are not interpreted as providing sufficient and necessary conditions. Instead, they may express defeasible connections between some natural properties and safety. Natural properties may in this sense perfectly well be indicative of certain risk and safety properties. 'When there is smoke, there is fire' is a perfectly sensible statement. Few would interpret this as a claim that smoke necessarily means fire. Similarly, even if we should refrain from reducing risk to the expectation value of harm, the expectation value surely does a good job of indicating risk. If the expected value of harm for A is much higher than for B, that normally signifies that B is safer than A, and there needs to be some argument present for not drawing that conclusion. However, if an argument is forthcoming, the inference may be blocked.

What does follow from my argument is that natural science is not the indefeasible determinant of risk and safety. Hence, risk analysis must be understood as an open process in an important sense. While the scientific method is the best out there to capture the natural basis that typically indicates risks at hand, the normativity of risk and safety means that the validity of the natural science indication may always be questioned. The process of risk analysis should thus be dialectical: in principle at least, it should always be open for the possibility that the natural conceptualisations and operationalisations used are insufficient. In practice, this implies that there should be processes not only for actively including alternative perspectives for evaluation but also for handling new justificatory claims. In other words, it should be possible to raise arguments that question that the right sort of considerations for the situations at hand have been included into the assessment and managed in the risk process. Practical limitations may of course restrict this ideal process, but in principle, the best argument, including the best understanding of the salient features of the current situation, should be able to decide the outcome.

In conclusion, codifying indicators for comparing risks are valuable and practically indispensable, just as rules of thumb in other areas. Here the scientific experts play an indispensible role. But in our striving for practical operationalisations, we should not forget that the extent to which our answers are good does not only depend on how easily we may find them.[33] Questions of risk and safety are intimately linked to questions of what to do, and here we should not expect any easy way out, especially not in a naturalistic form.

References

Ackerman, F., and L. Heinzerling. 2002. Pricing the priceless: cost-benefit analysis of environmental protection. *University of Pennsylvania Law Review* 150: 1553–1584.

Aven, T., and V. Kristensen. 2005. Perspectives on risk: Review and discussion of the basis for establishing a unified and holistic approach. *Reliability Engineering and System Safety* 90: 1–14.

Ayer, A. 1936/2001. *Language, truth and logic*. London: Penguin Books.

Blackburn, S. 1984. *Spreading the word*. Oxford: Clarendon Press.

Blackburn, S. 1998. *Ruling passions: A theory of practical reasoning*. Oxford: Clarendon Press.

Cohen, B.L. 2003. Probabilistic risk analysis for a high-level radioactive waste repository. *Risk Analysis* 23: 909–15.

Dancy, J. 1995. In defence of thick concepts. In *Midwest studies in philosophy 20*, ed. French, Uehling, and Wettstein. University of Notre Dame Press, Notre Dame, Ind.

Dancy, J. 2004. *Ethics without principles*. Oxford: Clarendon Press.

Douglas, M., and A. Wildavsky. 1982. *Risk and culture: An essay on the selection of technological and environmental dangers*. Berkeley: University of California Press.

EC. 2003. Technical Guidance Document in support of Commission Directive 93/67/EEC on Risk Assessment for new notified substances, Commission Regulation (EC) No 1488/94 on Risk Assessment for existing substances and Directive 98/8/EC of the European Parliament and of the Council concerning the placing of biocidal products on the market, Luxembourg: Joint Research Centre, EUR 20418 EN, Office for Official Publications of the EC.

Elgin, C. 2005. Williams on truthfulness. *The Philosophical Quarterly* 55: 343–352.

Elstein, D., and T. Hurka. 2009. From thick to thin: Two moral reduction plans. *Canadian Journal of Philosophy* 39: 515–536.

Hansson, S.O. 2004. Philosophical perspectives on risk. *Techne* 8: 1.

Hansson, S.O. 2005. Seven myths of risk. *Risk Management* 7: 7–17.

Hansson, S.O. 2010. Risk: Objective or subjective, facts or values. *Journal of Risk Research* 13: 231–238.

Hare, R.M. 1952. *The language of morals*. Oxford: Clarendon.

Hare, R.M. 1997. *Sorting out ethics*. Oxford: Clarendon.

Hooker, B., and M. Little. 2000. *Moral particularism*. Oxford: Clarendon Press.

Kraus, N., T. Malmfors, and P. Slovic. 1992/2000. Intuitive toxicology: Experts and lay judgements of chemical risks. In Slovic, 285–315.

Leiss, W. 2004. Effective risk communication practice. *Toxicology Letters* 149: 399–404.

Little, M. 1994. Moral realism II: Non-naturalism. *Philosophical Books* 35: 225–232.

Little, M. 2000. Moral generalities revisited. In Hooker and Little.

Mackie, J. 1977. *Ethics: Inventing right and wrong*. Harmondsworth: Penguin.

McDowell, J. 1978. Are moral requirements hypothetical imperatives? *Proceedings of the Aristotelian Society Supplementary* 52: 13–29.

[33] In the words of W. D. Ross: "Loyalty to the facts is worth more than a symmetrical architectonic or a hastily reached simplicity." (Ross (1930/1967), 23.)

McDowell, J. 1979. Virtue and reason. *Monist* 62: 331–350.
McDowell J (1981) Non-cognitivism and rule-following. In Wittgenstein: To follow a rule, ed. S. Holtzman and C. Leich, 141–162. London/Boston: Routledge & Kegan Paul.
McNaughton, D., and P. Rawling. 2000. Unprincipled ethics. In Hooker and Little, 256–275.
Miller, A. 2003. *An introduction to contemporary metaethics*. Oxford: Polity Press.
Möller, N. 2009. Should we follow the experts' advice? Epistemic uncertainty, consequence dominance and the knowledge asymmetry of safety. *International Journal of Risk Assessment and Management* 11: 219–236.
Möller, N. 2011. The concepts of risk and safety. In *Handbook of risk theory*, eds. S Roeser, R. Hillerbrand, M. Peterson, and P. Sandin. Springer.
Möller, N., S.O. Hansson, and M. Peterson. 2006. Safety is more than the antonym of risk. *Journal of Applied Philosophy* 23: 419–432.
Moore, G.E. 1903. *Principia ethica*. Cambridge: Cambridge University Press.
National Research Council. 1983. *Risk assessment in the federal government managing the process*. Washington, DC: National Academy Press.
Putnam, H. 1975. The meaning of 'Meaning'. In *Mind, language and reality: philosophical papers*, vol. 2, 215–271. Cambridge: Cambridge University Press.
Rawls, J. 1971. *A theory of justice*. Cambridge, MA: Belknap Press of Harvard University Press.
Ross, W.D. 1930/1967. *The right and the good*. Oxford: Clarendon Press.
Sandin, P., M. Peterson, S.O. Hansson, C. Rudén, and A. Juthe. 2002. Five charges against the precautionary principle. *Journal of Risk Research* 5: 287–299.
Slovic, P. 2000. *The perception of risk*. London: Earthscan.
Stevenson, C. 1944. *Ethics and language*. New Haven: Yale University Press.
Vallentyre, P. 1998. The nomic role account of carving reality at the joints. *Synthese* 115: 171–198.
Wikman-Svahn, P., M. Peterson, and S.O. Hansson. 2006. Principles of protection: A formal approach for evaluating dose distributions. *Journal of Radiological Protection* 26: 69–84.
Williams, B. 1985. *Ethics and the limits of philosophy*. Cambridge, MA: Harvard University Press.
Wynne, B. 1992. Carving out science (and politics) in the regulatory jungle. *Social Studies of Science* 22: 745–58.

Chapter 12
Risk and Degrees of Rightness

Martin Peterson and Nicolas Espinoza

1 Introduction

Suppose you accept a moral norm according to which it is wrong to harm a person. How should you then reason about a case in which someone will be harmed with a probability less than one? How exactly are we to draw the line between morally acceptable risks and risks that should not be accepted? The aim of this chapter is to suggest a novel answer to this question, which we believe can be rendered compatible with both consequentialist and non-consequentialist positions in normative ethics.

Briefly put, we hypothesize that some risky actions are best conceived of as being right or wrong to some degree. That an action is right to a certain degree means that it is more right (or righter) than an action that is right to a lower degree. From a philosophical perspective, this hypothesis is novel, and it has important implications for how we should think about the structure of normative ethics.[1] Existing views about risk are all based on the assumption that there is a sharp line to be drawn between risks that are morally acceptable and risks that are not. According to the hypothesis, we propose this is however an overly simplistic structural limitation. By acknowledging that basic moral categories such as rightness and wrongness come in degrees, we can provide a more nuanced picture of the ethics of risk.

The basic idea can be formulated as follows. Suppose you accept a moral theory *m* according to which it is morally wrong to bring about a state *s* with probability one.

[1] For a discussion of how how this proposal can be rendered compatible with consequentialist moral theories, see Peterson (2013).

M. Peterson (✉)
Eindhoven University of Technology, Eindhoven, the Netherlands
e-mail: m.peterson@tue.nl

N. Espinoza
Stockholm University, Stockholm, Sweden
e-mail: nicolas.espinoza@philosophy.su.se

M.J. de Vries et al. (eds.), *Norms in Technology*, Philosophy of Engineering and Technology 9, DOI 10.1007/978-94-007-5243-6_12,

Then, on our view, there is some probability p strictly lower than one, such that according to m, bringing about s is right to some (non-maximal) degree and wrong to some (non-maximal) degree. However, the claim that risky actions are right and wrong to some degree does not commit us to any particular view about what exactly make right actions right and wrong ones wrong. We primarily take our hypothesis to be a claim about the structure of normative ethics, rather than a first-order normative claim. If accepted, it enables the moral theorist to distinguish between a wide range of possible moral verdicts about risk, for example, "entirely wrong," "wrong to some degree," and "almost entirely right." The biggest challenge that lies ahead of us is, presumably, to explain what it means to say that rightness and wrongness vary in degrees.

The structure of this contribution is as follows. In Sect. 2, we summarize some traditional consequentialist views about risk and discuss some of the main objections against them. In Sect. 3, we discuss Nozick's rights-based view, which is among the most interesting and detailed non-consequentialist discussions of the ethics of risk. In the remaining sections, Sects. 4, 5 and 6, we outline and defend our own degree-based account.[2]

2 Consequentialism and Risk

Consider a set of three alternative actions, A, B, and C. Depending on which state of the world, s or t, happens to be the true state and which action is chosen, exactly one of the following six possible outcomes will materialize.[3]

The numbers in Table 12.1 represent utility, that is, a measure of value that is invariant up to positive linear transformations. Actualist accounts of consequentialism maintain that one ought to perform A if and only if s is the true state of the world; otherwise, B ought to be performed. This holds true no matter whether the agent can foresee which state will become the actual one. Actualism is a criterion of rightness, not a decision-making procedure.[4]

Many consequentialists find the actualist doctrine normatively unacceptable. Although it might be reasonable to claim that it is better that an optimal outcome is produced than a nonoptimal one, it is a mistake to identify the deontic status of an action with its actual consequences.[5] Consider the choice between A and B. If you choose B, it is certain that no really good or really bad events will occur. However, if you choose A, there is some probability p that the consequences will be very bad—but there is also some probability $1—p$ that the consequences will be very

[2] Sections 5 and 6 of this chapter draw heavily on a forthcoming paper by Espinoza and Peterson called "The Degrees of Moral Rightness."

[3] For a similar example, see Jackson (1991).

[4] The *locus classicus* is Bales (1971).

[5] The term deontic status refers to whether the action is right or wrong, permissible or obligatory, and so on.

Table 12.1 Consequentialism and risk

	State *s*	State *t*
Alternative A	100	−100
Alternative B	−10	10
Alternative C	1	1

good. Let us suppose that you know that the probability of producing the very good consequences (by performing A) is just one in a million, but that you nevertheless decide to do A. Let us furthermore suppose that the actual consequences of performing A turn out to be very good. Intuitively, many people now feel that the decision to perform A was irresponsible and ought not to have been taken—it was far too risky. If we take our moral intuitions seriously, we should arguably perform B rather than A. This is incompatible with the recommendations elicited by actualist theories.

Moreover, even if moral rightness was to depend exclusively on how much utility is actually produced, it does not follow that there is always a sharp line to be drawn between right and wrong actions. In order to demonstrate this point, it suffices to consider cases of evaluative incomparability: Suppose you could either (1) make 10,000 people a little bit happier for a short period of time by writing a funny text for your university newsletter or (2) make a single prisoner much happier for several decades by writing a debate article that convinces the authorities that he is innocent and therefore has to be released from prison.

Some traditional actualists would presumably argue that some action is always certain to produce at least as much utility as all others, that is, that "as least as much utility as" is a complete ordering. The problem with this escape route is, however, that it is vulnerable to the small improvement argument. The small improvement argument goes as follows: Suppose that the number of people who will read your funny text is so large that neither of the two options will produce more utility than the other. Does it follow that both actions will therefore produce exactly as much utility? It seems that the answer is no. In order to see this, suppose that you were to write a column that is just a little bit funnier than the first one—that is, we assume that the first alternative can be improved just a tiny little bit. Then it does not follow that writing the slightly funnier column would as a matter of fact produce more utility than writing the debate article. These two options are very different, and the two kinds of utility we are considering (laughing, being released from prison) cannot be compared. It is not the case that one alternative produces more utility than the other, nor do they produce equally much utility. According to the Small Improvement Argument, this shows that "as least as much utility as" is an incomplete ordering, that is, that some outcomes are incomparable with respect to utility.

For the actualist who accepts incomparability, it seems that it is difficult to uphold a sharp distinction between right and wrong. If the underlying utility function is fuzzy, it seems that the corresponding moral categories could very well come in degrees. To insist on a binary distinction between right and wrong appears arbitrary. We propose that risk could be treated in ways analogous to cases involving incomparable utilities.

Many consequentialists disagree with the actualist view without thinking that moral rightness comes in degrees. This is by no means incoherent. It is certainly possible to insist on taking other features than the actual outcome into account when specifying a criterion of rightness, without introducing a degree-based account. For example, many scholars argue that consequentialists should rather take the expected consequences to constitute the basis of a reasonable criterion of rightness (see, for example, Brandt 1959; Jackson 1991; Smart 1973). Hence, if states s and t in Table 12.1 were to be equiprobable, it would follow that the agent is morally obliged to perform action C.

Again, it might of course be objected that the expected utility principle presupposes that there are no incomparable utilities. If such utilities exist, the traditional expected utility principle can only produce sharp moral verdicts by imposing binary borders on phenomena for which there are none to be found.

An additional and perhaps more severe problem with the expected utility criterion is that it yields unintuitive recommendations in many cases. Since the expected utility criterion does not take the variance of the utility distribution into account, it follows that an action that produces a very good or a very bad outcome with equal probabilities has exactly the same deontic status as an action that is certain to produce a mediocre outcome.

All these objections are familiar from the literature (see, for example, Smart 1973). We do not claim to have added anything new. However, by reconsidering these familiar problems with traditional consequentialist analyses of risk, it is easier to appreciate the advantages of our novel hypothesis, that is, the claim that risky actions are sometimes right and wrong to some degree and that binary accounts of rightness and wrongness are therefore inappropriate. Applied to the example in Table 12.1, the idea would be as follows: If the probability of state s is p and the probability of state t is $1-p$, then it is right to degree p to perform A and right to degree $1-p$ to perform B. Why? Because A would be optimal if the state s obtains (ten units of utility would be gained), and the probability of state s is p. However, if state t were to obtain, more utility would be produced by performing B, and the probability of this state is $1-p$. Hence, it is right to degree $1-p$ to choose alternative B.

3 Rights and Risk

Robert Nozick is one of the few non-consequentialists who acknowledges that the prevalence of risk raises fundamental ethical concerns. Nozick writes that "Actions that risk crossing another's boundary pose serious problems for a natural-rights position... Imposing how slight a probability of a harm that violates someone's rights also violates his rights?" (Nozick 1974, p. 74). It would arguably be implausible to claim that it is always wrong to impose involuntary risks on others; then nearly all actions would be wrong, such as driving to work or cooking dinner on your stove for your family. Moreover, to just pick a fixed numerical probability and then use it for distinguishing between morally acceptable and unacceptable risks seem arbitrary.

In order to make a long story short, Nozick concludes that risk impositions are morally permissible if and only if people imposing risk on others "compensate those disadvantaged through … these risky activities" (Nozick 1974, p. 87). The size of the compensation should be determined by the loss incurred by the risk. If the loss incurred by the risk is large, then the compensation has to be large, and vice versa. This means that Nozick is not an actualist about risk: It is not just the actual outcome that matters. Even a mere possibility that you may harm someone is morally relevant, even if no one is actually harmed.

On Nozick's view, the risk of physical harm is itself a special kind of moral harm, which entitles the victim to compensation. Suppose, for instance, that an airline fails to maintain its aircraft properly and that a large number of passengers therefore run an increased risk of being killed. Since the passengers have good reasons to suppose that the aircraft fleet is being maintained properly, this is not a risk to which they automatically consent by buying a ticket. Even if no aircraft ever crashes and no one is hurt, the increased involuntary risk counts as a harm that entitles the passengers to some compensation.

But how do we determine how much compensation is required? According to Nozick, "Something fully compensates a person for a loss if and only if it makes him no worse off than he otherwise would have been."[6] Therefore, if the airline exposes its passengers to some unforeseen and involuntary risk, the airline becomes morally obliged to pay the passengers an amount of money (or provide services valued by the passengers) that makes the passengers exactly as well off as they would have been in case they would not have been exposed to the risk. In Nozick's view, the correct way of establishing whether something makes someone better or worse off than he or she was before is to allow people to buy and sell commodities on a free market. If you are willing to swap x for y, this transaction will leave you no worse off than before. In this case, the commodities traded on the market are risks. The buyers of risks are usually insurance companies, and the sellers are individuals exposed to risks they consider to be too large. This means that the premium you have to pay for buying an insurance that protects you against a certain risk corresponds to the amount of compensation that you are entitled to when being exposed to an involuntary risk.

An obvious problem with Nozick's position is that not all losses can be easily compensated for. If you are killed in an airplane accident, no amount of money, no matter how large, will compensate for this. According to Nozick, this is not a fatal objection to his view: "While alive, [the victim] can sell the right to this payment, should it have to be made, to a company that purchases many such rights. The price would be no greater than the right's expected monetary value."[7]

Another and more serious problem with Nozick's theory, which he does not comment on, is that his emphasis on compensation makes him vulnerable to objections usually

[6] *Ibid*. p. 57.

[7] *Ibid*. p. 77.

raised against consequentialist theories. For Nozick, it would in principle be permissible to impose any kind of harmful risk on non-consenting individuals (such as a 99% chance of being tortured) given that one also offers a sufficiently large compensation in return (such as a huge amount of money). This however seems to be the wrong conclusion, at least if one believes that rights play a fundamental role in ethics. In order to see this, suppose that $20 million is sufficient compensation for a 99% chance of being tortured. It would then be morally permissible for you to impose this risk on your neighbor—as long as you give her $20 million—no matter whether she/he actually consents to being exposed to this risk or not. Just because the market price of an insurance against this kind of risk is $20 million, it does not follow that you would actually consent to being exposed to this risk. Not all individuals are willing to buy and sell goods on the market just because there is a market for the good in question. However, for moral theorists who take natural rights seriously, actual consent is what ultimately matters. You may only buy something from your neighbor if he or she actually agrees to your offer—the neighbor has the right to say no to the proposed transaction even if the price you offer is way above the market price. The morally relevant consideration is whether your neighbor is actually willing to sell the good you are bidding for. However, on Nozick's view, this does not apply to risk.

For consequentialists and other critics of Nozick, it is of course not implausible to claim that one need not always respect the autonomy of the individual. The overall consequences of accepting some involuntary risks are sometimes likely to be good. The problem is that Nozick cannot claim that one should accept involuntary risks because the consequences are good—that is not a relevant consideration for him. Indeed, the very idea of allowing people to impose some involuntary risks on others seems to be at odds with the spirit of a rights-based theory. Yet, as we have seen above, his theory allows for precisely this. The fact that people are compensated is irrelevant.

4 Degrees of Rightness

On our view, some risky actions are neither fully right nor fully wrong. We claim that such actions are rather right and wrong to some degree. Our claim about degrees of rightness can be rendered compatible with consequentialist as well as non-consequentialist accounts of risk. Consequentialists could claim that the probability of reaching an optimal outcome determines the degree to which the action is right, whereas the probability of reaching a suboptimal outcome determines the degree to which the action is wrong. The sum of these probabilities will always be one. In a similar way, people advocating natural rights could claim that the probability that all rights are respected determines the degree to which the action is right, whereas the probability that at least one right is violated determines the degree to which the action is wrong. Our view thus nicely captures the intuition that an action is more wrong the more probable it is that something morally undesirable occurs, and more right the more probable it is that something morally desirable occurs. There is no sharp boundary between morally acceptable and unacceptable risks.

Although never extensively discussed in the literature before, the idea that moral rightness varies in degrees has been briefly touched upon by several authors. John Stuart Mill claimed that, "The creed which accepts as the foundation of morals, Utility, or the Greatest Happiness Principle, holds that actions are right in proportion as they tend to promote happiness, wrong as they tend to produce the reverse of happiness" (Mill 1969, p. 210) If we interpret Mill literally, some acts are clearly more right than others. Eriksson argues that this feature of Mill's theory has been overlooked (Eriksson 1997). Eriksson also proposes his own version of utilitarianism, according to which, "There are not only discrete deontic qualities, such as rightness and wrongness, but within the space of wrong actions there is a continuum ranging from, say, very wrong indeed to nearly right."[8] Another author who has discussed similar ideas, and in somewhat more detail, is Ted Lockhart. In his Moral Uncertainty and Its Consequences, he considers but never explicitly defends the hypothesis that moral rightness and wrongness come in degrees (Lockhard 2000, Chapter 4). However, a major difference between Lockhart's account and ours is the following: On Lockhart's view, degrees of rightness can only occur if three or more options are compared; whenever two (different) options are compared, one option is always entirely right and another one entirely wrong. We do not agree with this, that is, we do not think that it is always possible to reduce complex moral decisions into pairwise comparisons between options that are entirely right or entirely wrong.

The claim that obligations can vary in degree should be kept apart from the idea that obligations can vary in strength. For example, your obligation to prevent someone's death is stronger than your obligation to prevent a headache. Sloman (1970) suggests that differences in strength can be captured by the difference between "ought" and "must"—you ought to prevent the headache, but you must prevent the death of your friend. Arguably, the strength of an obligation has to do with the amount of moral value at stake. Your obligation to prevent a death is stronger than your obligation to prevent a headache simply because there is less at stake, morally speaking, in the latter example. Clearly, this notion of strength does not coincide with the notion of degree we have in mind.

The view we propose, that the deontic status of actions can vary in degrees, goes against the standard view that actions are always either right or wrong. The standard view is accepted by both those who believe that conflicting obligations— such as the obligation to produce good outcomes and the obligation to minimize risks—can always be resolved into an all-things-considered obligation (which in turn makes some actions come out as right and the rest as wrong), and those who believe that conflicting obligations sometimes give rise to irresolvable moral dilemmas. These two positions, RESOLUTION and DILEMMA, are both committed to the claim that moral obligations are discrete entities that are either valid or not. On the RESOLUTION view, some available actions are right, and all other actions are

[8] *Ibid.*, p. 213.

wrong because there always exists a single all-things-considered obligation that holds to the highest degree. On the DILEMMA view, all available actions are wrong because there exists two or more all-things-considered obligations that hold to the highest degree. We argue, however, that by rejecting the standard view, we make room for a third position according to which one's all-things-considered obligation to perform one of two incompatible actions will in many cases be merely partial, that is, hold to a limited degree. Such obligations can be thought of as being brought about by an underlying structure of conflicting obligations, which only in part supports performing one of the actions and only in part supports performing the other. We will call this position DEGREE.

The three views introduced here—RESOLUTION, DILEMMA, and DEGREE—are pairwise incompatible. This is a definitional truth. However, it is not at all clear whether they are also jointly exhaustive. There seems to be some logical space left for views that reject all three alternatives. One such view, which we briefly touched upon above, holds that some actions have no deontic status whatsoever and that the remaining ones are either right or wrong. For simplicity, we shall treat this view as a special case of either DILEMMA or RESOLUTION, depending on whether it also allows for cases in which all available alternatives are wrong.

There is also logical space left for another view, which is not identical to any of those introduced so far and which has a substantial number of supporters among moral philosophers. On this view, some actions are supererogatory. That an action is supererogatory means that it is in some sense morally better than merely right actions: supererogatory actions go beyond and above what is morally required. Such actions are therefore neither right nor wrong, or right or wrong to some degree, so they cannot be accounted for within the three positions introduced thus far.

We of course acknowledge that supererogation might be an interesting moral phenomenon that raises important questions. Existing accounts of supererogation are, however, discrete: the boundary between merely right and supererogatory actions is taken to be sharp, which means that no action is supererogatory to some degree and right to some degree. However, as we see things, there is little reason why people who believe in supererogation must accept such a discrete position. If DEGREE is a plausible claim about rightness and wrongness, one can presumably also imagine cases in which some actions are both supererogatory to some degree and non-supererogatory to some other degree. Although the concept of supererogation will not be discussed in any detail in this book, we acknowledge that one could in principle treat non-discrete accounts of supererogation as special cases of DEGREE.

All that said, we believe it is important to also comment on a position that does not play a role in our theory. This will help avoid misunderstanding. The point we wish to make is that DEGREE is not identical with, and does not depend on, any claim about vagueness. By definition, a predicate is vague if and only if its extension lacks sharp boundaries. "Bald" is a standard example: for some persons *p*, the proposition "*p* is bald" seems to be neither true nor false. If DEGREE were a claim about vagueness, it would thus be the claim that some moral predicates, such as right and wrong, lack sharp boundaries. In the same way as some persons are neither bald nor not bald, it could be claimed that some actions are neither right not-right.

In order to see why the hypothesis that moral rightness and wrongness come in degrees does not coincide with, or is dependent on, the hypothesis that "right" and "wrong" are vague predicates, it is helpful to first note that a general characteristic of vagueness is that small changes have no big effects. If someone is bald and suddenly gets one extra strain of hair, that will not take the person across the border and make him non-bald. This observation can be used for spelling out an important difference between DEGREE and vagueness: advocates of DEGREE can accept the idea that a small change may sometimes have big moral effects, even if the change occurs in the "gray area" in which actions are neither right nor wrong. Consider, for instance, the thin boundary between racist jokes and morally uncontroversial ones. Sometimes, a small change to the tone in which a joke is told, or a small change to which words are emphasized, can spell the difference between a joke that is morally acceptable to a high degree and one that is racist and entirely wrong. For another example, consider the way in which a male professor treats a female student. Sometimes, a very small and subtle change of the professor's behavior can transform a normal everyday action into a sexist and morally wrong one. That is, a small change can transform an action from being entirely right to a being wrong to a high degree. This indicates that moral rightness and wrongness can come in degrees even in situations that do not involve vagueness.

Other differences between DEGREE and vagueness can be identified by considering specific theories of vagueness. The dominant approaches to vagueness are supervaluationist theories, degree of truth theories, and epistemic theories (see, for example, Williamson 1994; Keefe 2000; Smith 2009). According to the supervaluationist approach to vagueness, a sentence is determinately true if and only if it is true on all its sharpenings, determinately false if and only if it is false on all its sharpenings, and indeterminate (neither determinately true nor determinately false) if and only if it is true on some sharpenings but false on other sharpenings. So, for example, the sentence "John is bald or John is not bald" will come out as determinately true since for every (admissible) sharpening of "bald" John will be either bald or not bald according to that sharpening. Now, there are two reasons our theory should not be misinterpreted as a mere instantiation of the supervaluationist account of vagueness. First, the supervaluationist approach is meant to allow for truth value gaps, not degrees of truth, which, intuitively, is what is needed to account for our degree account. Second, there is some controversy as to whether the supervaluationist account can make sense of truth value gaps at all. According to Williamson, the theory is self-defeating: by accepting Tarski's T-schema, and the law of excluded middle, supervaluationists are committed to bivalence, which excludes the possibility of truth value gaps (Williamson 1994, p.162–163).

Next, in order to articulate the difference between DEGREE and degree of truth theories of vagueness, note that the notion of degrees we wish to articulate is not that of degrees of truth. On a degree theory of vagueness, predicates are assigned truth values ranging from 0 to 1, but such an approach would apparently not suit our purpose. For example, it might be entirely true (true to degree 1) that it is morally right to some degree to rescue the people aboard a sinking cruise liner even if the rescue mission is dangerous and unlikely to be successful. But this is not equivalent to claiming that it is true to some degree that it is entirely right from a moral point of view to rescue the people.

According to epistemic theories, vagueness is an illusion created by our limited epistemic abilities. Whenever a predicate appears to be vague, there are in fact always sharp boundaries that determine its applicability. That is, for every person *p*, the proposition "*p* is bald" is either true or false. It is just that in borderline cases, we have not yet, and perhaps never will, be able to determine whether *p* is bald or not. No matter whether this is a reasonable view about vagueness or not, it is easy to see that it is distinct from DEGREE. This is because according to epistemic theories, there are only two possibilities: every person is either bald or not. So in a moral context, the advocates of epistemic views would have to argue that every action is either right or not. However, according to DEGREE, this is not the case. Some actions are neither right nor wrong, no matter what the epistemic situation happens to be. Advocates of DEGREE do not maintain that all actions are right and wrong and that the intuitions about degrees are due to epistemic limitations. The notion of degrees that advocates of DEGREE believe in is a genuine, non-epistemic one—not a merely apparent one.

What we learn from the discussion above is that DEGREE may be a plausible hypothesis even if we know for sure that we are not facing a case of vagueness. DEGREE is compatible with the claim that small changes of morally relevant features can lead to big changes in the moral evaluation of an action. By just changing something a little bit, the degree to which an action is right can change significantly. Moreover, unlike vagueness, DEGREE is also compatible with the hypothesis that there are sharp boundaries between different degrees. The spectrum of degrees need not be a continuous one.

5 Our Argument for DEGREE

In this section, we outline our positive argument for accepting DEGREE. Our argument is based on two premises. The first is the moral claim that we sometimes have strong moral reasons for and against performing an action, and that in such cases, one's all-things-considered obligation to perform (or not to perform) the action can be valid to some limited degree. For example, the fact that a risky action may harm someone or violate someone's rights indicates that one has an obligation not to perform the risky action. Moreover, the fact that the risk may never materialize, while the action is necessary for realizing some other moral goal (such as optimal consequences, or respecting some rights), indicates that one also has an obligation to perform the action in question. Let us call this premise conflicting obligations (CO). Scholars advocating DILEMMA can easily accept CO. According to them, the best explanation of why all alternative actions are sometimes (entirely) wrong is that our obligations clash. However, in contrast to scholars advocating DILEMMA, supporters of RESOLUTION are keen to reject CO. This is because they are committed to the claim that at least one action is always (entirely) right; it is therefore tempting to maintain that what appears to be a clash between incompatible moral obligations will invariably, at the end of the day, turn out to be a conflict between evidential obligations that can be resolved into a set of nonconflicting moral obligations.

That said, it should be stressed that CO is logically independent of DILEMMA and RESOLUTION. Both these positions are claims about the deontic status of actions, whereas CO is a claim about moral obligations. It is logically possible, but perhaps not very attractive, to maintain that the deontic status of an action is independent of or compatible with any moral obligations obtaining in a situation. Obligations and actions are different entities. In order to justify DEGREE, we therefore need to somehow connect one's moral obligations with the deontic status of actions.

This naturally brings us to our second premise. The second premise holds that the deontic status of an action corresponds to the moral obligations that obtain in the situation. To be more precise, the premise holds that an action is right to a (non-extreme) degree if and only if one has a moral obligation to some degree to perform it and a moral obligation to perform some other action that is incompatible with the action in question. Moreover, an action is entirely right if and only if one has at least one moral obligation to perform it and no moral obligation not to perform it, and it is entirely wrong if one has at least one moral obligation not to perform it and no moral obligation to perform it. One may thus think of the second premise as a Correspondence Principle (CP): the deontic status of an action corresponds to the obligations that determine its deontic status.

Our argument for DEGREE is then, briefly put, that the first-order premise CO in conjunction with the correspondence principle CP supports the conclusion that moral rightness and wrongness vary in degrees. In summary:

CO: Sometimes (e.g., in situations involving risk) one has a moral obligation to perform an action and a moral obligation to perform some other action that is incompatible with the action in question.

CP: The deontic status of an action corresponds to the moral obligations obtaining in the situation.

DEGREE: Some risky actions are right to some (non-extreme) degree.

DEGREE is not logically entailed by CO and CP. The support offered by the two premises is nevertheless significant. Premises CO and CP give a clear overview of what the debate over RESOLUTION, DILEMMA, and DEGREE is all about. If you accept both premises, the best explanation of this is arguably that you accept DEGREE. The challenge that lies ahead of us is, therefore, to first overcome the somewhat empty formalism of CO and CP by filling their central terms with philosophical content, and thereafter defend these claims from a moral point of view.

The purpose of the reminder of this chapter is primarily to defend the second premise, CP. We believe that premise CO has already been sufficiently supported. In what follows, we shall thus simply take for granted that CO is true. CP is a claim about how certain concepts are interrelated. It is thus partly a conceptual claim, not a pure first-order moral claim. However, our defense of CP is not based on any argument about how people do actually use concepts such as "moral obligation," "right," and "wrong." We do not think that semantic observations of how people actually use very complex and abstraction concepts are sufficient for informing us about how these concepts ought to be used. In what follows, we will rather argue that by extending the traditional concept of rightness (and wrongness) from a binary concept to a degree-based concept, we can obtain a more precise and morally plausible analysis of a number of moral situations. So our argument for revising the traditional binary concept is a pragmatic one: If we go ahead along the suggested path, we will be able to acquire a conceptual apparatus that better serves the purposes of moral philosophy.

6 The Correspondence Principle

The best reason for accepting CP is that it is required for avoiding deontic leaps. The term "deontic leap" is a technical term we use for describing the process in which a moral theory assigns some deontic status to an action that does not correspond to the moral obligations present in the situation.

Imagine, for instance, that someone claims that it is entirely right, all-things-considered, to break a promise in a situation in which two conflicting moral obligations are in force, viz., an obligation not to break a promise (because promises should always be kept) and an obligation to break the promise (because the consequences would be optimal). This would be an extreme example of mismatch between the deontic status of an action and the moral obligations in play. By accepting CP, we can prevent such unwarranted deontic leaps from arising.

Deontic leaps are analogous to a kind of unjustified leaps sometimes discussed in epistemology. Let us, for illustrative purposes, consider the epistemic leap that drives Kyburg's infamous lottery paradox. This paradox can be formulated by considering a fair 1,000 ticket lottery that has exactly one winning ticket. If the rules of the lottery are known by the player, it is therefore rational for her/him to accept the proposition, "some ticket will win." However, also suppose that an event is very likely only if the probability of it occurring is greater than 0.99. On these grounds, it is presumed rational to accept the proposition that ticket 1 of the lottery will not win (since one should not accept propositions that are very unlikely to be true). Since the lottery is fair, it is rational to accept that ticket 2 won't win either; indeed, it is rational to accept for any individual ticket i of the lottery that ticket i will not win. However, accepting that ticket 1 will not win, and accepting that ticket 2 will not win, and so on, seems to entail that it is rational to accept that no ticket will win, which contradicts the conclusion above that some ticket will win.

An influential resolution of the lottery paradox, most notably affirmed by Jeffrey (1970), is to argue that we should reject the idea of outright belief altogether.[9]

[9] We are aware that there are other competing solutions to this paradox but mention Jeffrey's solution for illustrative purposes.

According to Jeffrey, "Ramsey sucked the marrow out of the ordinary notion [of belief], and used it to nourish a more adequate view," which of course is the view that belief comes in degrees and can be measured by a subjective probability function. So on this view, the appropriate degree of belief that ticket i will not win is 0.999. To accept that ticket i will not win is to make an unjustified epistemic leap from 0.999 to 1.

That said, there are of course other ways of resolving Kyburg's lottery paradox. The most obvious proposal is the following: Although a rational agent ought to believe, and accept, that it is very likely that ticket i will not win, it does not follow that one ought to believe, and accept, that ticket i will not win. The move from a claim about what is very likely to happen to a claim about what will happen is unjustified. That is, outright beliefs should not approximate the evidence; a rational epistemic agent should rather hold outright beliefs that are fully supported by the evidence. In the lottery paradox, the only fully supported belief is the belief that it is very likely that ticket i will not win. So rather than believing to a certain degree that *p* will occur, one should believe (fully) that it is likely that *p* will occur. However, such an alternative resolution of the lottery paradox seems to be difficult to apply in a moral context. Although one could perhaps claim that an action is right to a certain degree if two or more moral obligations clash, it seems to make little sense to claim that it could ever be "entirely right to perform an action to a limited degree." One can believe to a high degree that something will happen, and one can believe fully that something is very likely to happen, but it seems that in moral contexts, the analogous distinction makes little sense.[10]

Unless we accept the correspondence principle (CP), unwarranted deontic leaps are bound to arise. Imagine, for instance, a person who is about to choose between two alternative actions. Depending on which action is chosen and which of two equiprobable states of the world happens to be the true one, exactly one of four possible outcomes will materialize. If alternative D is chosen, 10 lives will be saved with 50% probability and zero lives with 50% probability. However, if alternative E is chosen, four lives will be saved with certainty.

As pointed out in Sect. 2, traditional act utilitarians maintain that it is right to perform alternative D if and only if it turns out that five lives are actually saved; otherwise, alternative E is right. Other consequentialists argue that one should rather take the expected consequences to constitute the basis of a reasonable criterion of rightness. Hence, since the two possible outcomes of alternative D are equiprobable, it is right to perform D because its expected utility is five, which is one unit more than the expected utility of E.

The claim that risk may make moral rightness and wrongness vary in degrees does not coincide with any of the traditional consequentialist views, nor with any other major ethical perspective. To say that one is obliged to perform both actions,

[10] Perhaps the distinction could be upheld in a moral context by introducing degrees of truth. It may very well be *entirely true* that it is obligatory to some degree to rescue the people aboard a sinking cruise liner even if this is dangerous, but this is not equivalent to saying that it is true to some degree that it is *entirely obligatory* to rescue the people.

to a certain degree, seems to tally well with our considered moral intuitions. For the consequentialist, what speaks in favor of alternative D is that this option may lead to the best outcome, whereas alternative E guarantees a higher minimum level. Arguably, this is a relevant moral consideration, which one should expect to affect our moral obligations. Furthermore, unlike advocates of actualist accounts of utilitarianism, people defending this mixed view need not insist on a razor-sharp distinction between normative and epistemic issues. If one accepts the claim that one is obliged to a certain degree to perform both D and E, for the reasons outline above, it follows that one will in nearly all cases be able to know this. Objections to actualist versions of consequentialism based on epistemic considerations do not seem to apply to DEGREE. Finally, as pointed out above, DEGREE can also be rendered compatible with rights-based moral theories.

References

Bales, R.E. 1971. Act-Utilitarianism: Account of right-making characteristics or decision-making procedure? *American Philosophical Quarterly* 8: 257–265.

Brandt, R. 1959. *Ethical theory: The problems of normative and critical ethics*. Englewood Cliffs: Prentice-Hall.

Eriksson, B. 1997. Utilitarianism for Sinners. *American Philosophical Quarterly* 34: 213–228.

Jackson, F. 1991. Decision-theoretic consequentialism and the nearest and dearest objection. *Ethics* 101: 461–482.

Jeffrey, R. 1970. Acceptance vs. Partial Belief. In *Induction, acceptance, and rational belief*, ed. M. Swain. Dordrecht: Reidel.

Keefe, R. 2000. *Theories of vagueness*. Cambridge: Cambridge University Press.

Lockhard, T. 2000. *Moral uncertainty and its consequences*. Oxford: Oxford University Press.

Mill, J.S. 1969. *Utilitarianism, The Collected Works of John Stuart Mill*, vol. X. Toronto: University of Toronto Press.

Nozick, R. 1974. *Anarchy, State, Utopia*. New York: Basic Books.

Peterson, M. 2013. *The dimensions of consequentionalism*. Cambridge: Cambridge University Press.

Sloman, A. 1970. `Ought' and `Better'. *Mind* 79(315): 385–394.

Smart, J.J.C. 1973. An outline of a system of Utilitarian Ethics. In *Utilitarianism: For and against*, ed. J.J.C. Smart and B. Williams. Cambridge: Cambridge University Press.

Smith, J.J. 2009. *Vagueness and degrees of truth*. Oxford: Oxford University Press.

Williamson, T. 1994. *Vagueness*. London: Routledge.

Chapter 13
Naturalness, Artefacts, and Value

Per Sandin

1 Introduction

Most of us have an intuitive understanding of the ontology of artefacts – we have a grasp of what an artefact is. An archetypal technical artefact consists of some tangible physical stuff that someone has modified to make it useful for some practical purpose – such as when wood and metal are treated and shaped to make a hoe, or when a piece of plastic is molded to make a canoe. I believe that this intuitive understanding is neatly captured by the following definition of (technical) artefacts proposed by Peter Kroes and Anthonie Meijers: They are "(i) designed physical structures, which realize (ii) functions, which refer to human intentionality" (Kroes and Meijers 2006, p. 2).

Given (ii), it is obvious that such artefacts can have instrumental value. A good cheese grater is one which does the job of grating the cheese efficiently, comfortably, etc. Some cheese graters are better than others – the pyramid grater is superior to the flat grater that consists of only a single plate. (I am aware of the potential complication of whether one can speak of good gas chambers or torture equipment, but I will not discuss it here.)

A different question, however, is whether artefacts can have intrinsic or non-instrumental value. There is a recent discussion within the field that has been termed "analytic philosophy of technology" (Franssen et al. 2009) concerning artefacts and normativity. Analytic philosophy of technology is primarily rooted in philosophy of science. But of course, this is only one area in which considerable intellectual efforts have been devoted to the study of artefacts, technical as well as non-technical ones, such as works of art. Scholars in these areas have pondered somewhat different problems than those that have been most relevant from the perspective of philosophy of science or (analytic) philosophy of technology.

P. Sandin (✉)
Swedish University of Agricultural Sciences, Uppsala, Sweden
e-mail: per.sandin@slu.se

M.J. de Vries et al. (eds.), *Norms in Technology*, Philosophy of Engineering and Technology 9, DOI 10.1007/978-94-007-5243-6_13,

When it comes to the questions of artefacts and value, it might be worthwhile to look outside the areas that hitherto have been most thoroughly explored by analytic philosophers of technology. I will therefore turn to the field of environmental ethics, a field in which artefacts, nature, and normativity have played important roles. My purpose is, first, to introduce the philosophy of artefacts as it has been discussed in environmental ethics; secondly, to argue for the relevance of the idea of the "natural" in relation to artefacts. I will argue that artefacts can display differing degrees of naturalness. I will then discuss whether artefacts that are "more natural" are somehow preferable to "less natural" ones. In the course of doing this, I will provide a critical discussion on some recent versions of the naturalness argument. In the end, I will argue that if natural objects can possess intrinsic value, so can artefacts. To show this, I offer an adaptation of Richard Routley's "Last Person" argument, as developed by Keekok Lee. A general observation is that environmental ethics and analytic philosophy of technology can benefit from each other's insights in the analysis of artefacts and normativity.

2 Environmental Ethics and Philosophy of Technology

2.1 What Is Environmental Ethics?

The term "environmental ethics" can mean several things. Two main meanings of the term, however, can be identified. According to one of them, environmental ethics is the ethics of man's relationship to the environment. According to the other, the phrase "environmental ethics" is roughly synonymous to "non-anthropocentric ethics". The former characterization refers to the object of study, while the latter implies a particular axiological commitment. This normative commitment might take several forms. At its core, however, is the thesis that there can be value in nature, without human valuers. This might be expressed in different forms: That non-human nature is intrinsically valuable, that it possesses noninstrumental value, and so on.

2.2 Environmental Ethicists on Technology

Environmental ethicists and their relationship to technology is a complicated matter. However, if one judges from the published academic literature, it appears that many card-carrying environmental ethicists are anything but enthusiastic about technology.

Some seem downright hostile and side with the Heideggerian critique of technology (Strong 1992). Others, perhaps a majority, while professing not to be hostile to technology, argue that present-day technology is morally problematic and needs to be subject to evaluation, checks, and possibly reform. This, for instance, appears to have been ecosopher Arne Naess' position.[1] In his words, "the objectives of the deep ecological movement do not imply any depreciation of technology or industry, but they imply *general cultural control* of developments" (Naess 1989, p. 102, my emphasis). Naess also proposes a checklist for ecosophical technology assessment. The items on the list are questions that are supposed to contribute to the assessment of whether a particular technique expresses an improvement. Some items on the list are the following questions: "Does it [the technique] strengthen cooperation and harmonious togetherness with other workers? (…) Does the technique pollute directly or indirectly? (…) Does it promote equality or class differences at the place of work or more generally?" (Naess 1989, p. 95f).

The concluding words of what must be considered one of the prime canonical texts of the environmental ethics literature, Aldo Leopold's "The Land Ethic," provide an apt illustration of this "control approach" to technology:

> We are remodeling the Alhambra with a steam shovel, and we are proud of our yardage. We shall hardly relinquish the shovel, which after all has many good points, but we are in need of gentler and more objective criteria for its successful use (Leopold [1949] 1987, p. 226).

Let us rest for a moment with these words. I do not think that it is a coincidence that Leopold uses a distinctly technological artefact in his metaphor: the steam shovel. Leopold apparently cherished the metaphor. He had used it as early as 1933, in his John Wesley Powell Lecture delivered to the Southwestern Division of the AAAS (Leopold 1933, p. 637).[2] Then, however, he used the metaphor without the acknowledgement that the shovel – i.e., technology – "has many good points." In 1933, we were remodeling the Alhambra with a steam shovel – full stop. Still, the early text oscillates between admiration and deep skepticism about technology.

We of the machine age admire ourselves for our mechanical ingenuity; we harness cars to the solar energy impounded in carboniferous forests; we fly in mechanical birds; we make the ether carry our words or even our pictures. But are these not in one sense mere parlor tricks compared with our utter ineptitude in keeping land fit to live upon? (Leopold 1933, p. 636).

The skepticism, however, also runs deep. Leopold writes about "the economic isms" – socialism, communism, and fascism – as "apostles of a single creed: *salvation by machinery*" (Leopold 1933, p. 639, italics in original). Incidentally, he lumps those together with "the late but not lamented technocracy," arguing that they "outdo even Capitalism itself in their preoccupation with one thing: The distribution of more machine-made commodities to more people" (ibid.). It might be remembered against what backdrop Leopold wrote: He delivered his AAAS Lecture on May 1, 1933, almost to the day 3 months after Hitler's being appointed German chancellor.

---

[1] As developed in Naess (1989), Ch. 4.

[2] Leopold reuses large parts of the 1933 text in "The Land Ethic." He had been thinking and writing about some of the issues covered in the essays long before that, in unpublished work. See, e.g. Norton (1996).

Using the steam shovel as a metaphor for technology and its impact on nature – at least in the first half of the twentieth century – is certainly apt. Steam powered the industrial revolution, and steam shovels were widely used in the industrial transformation of the natural landscape in the form of railways during the latter half of the nineteenth century and perhaps even more notably, the construction of the Panama Canal (Stueland 1994), completed less than two decades before Leopold's "Conservation Ethic" appeared. The other part of the metaphor is also interesting. The Alhambra is a distinctly human creation – an artefact – and it is produced using technology, albeit a very different form of technology than the one that produced the steam shovel.

It might not be surprising that Naess and Leopold were both skeptical and ambivalent. However, one general observation is that with the exception of some technologies (for instance, genetic modification), environmental ethicists seem to have been less interested in technology than one might have expected. There are of course a number of eminent thinkers – like Andrew Light, for instance – who engage in philosophy of technology and environmental ethics with equal brilliance and who explore the intersection between the fields.

However, some of the themes environmental ethicists have discussed are certainly relevant for a normative philosophy of artefacts. In the remainder of the chapter, I will focus on this issue and in doing so I hope to illustrate some points that philosophers of technology might find valuable in environmental ethics. No doubt, there are numerous things that environmental ethicists could and should learn from philosophy of technology. It is my firm belief that the fields could cross-pollinate each other. In this present chapter, however, I will be primarily concerned with one of these directions.

Among other things, environmental ethicists have been concerned with the question whether nature, in the sense of non-human nature, has intrinsic or non-instrumental value, or whether humans are the only source of such value. The latter position is known as anthropocentrism. In the late 1960s, environmental ethicists started to criticize and question the anthropocentrism of mainstream ethical theory (White [1967] 2009; Sylvan [1973] 2009). The discussion has since branched off into many directions; some question its relevance while it is at the center of others' positions (see, e.g., McShane 2007).

3 The Value of Nature

An offspring of the discussion about nature and its intrinsic value is precisely the debate about the normative status of natural items in relation to man-made ones.

Before we take on that debate, however, we must note that the term "nature" has a number of meanings. Two of the main ones were identified by John Stuart Mill in his classic essay "Nature":

> In one sense, [nature] means all the powers existing in either the outer or the inner world and everything which takes place by means of those powers. In another sense, it means, not

everything which happens, but only what takes place without the agency, or without the voluntary and intentional agency, of man. (Mill [1874] 1988, p. 8)

Since Mill, a number of other senses of the term have been discussed in the literature. Lee, for instance, lists no less than seven, including Mill's two (Lee 1999, p. 82ff).[3] However, for the purposes of the present essay, I will stick to the idea of something's being natural, expressed in Mill's second sense, that of not being produced by humans. Lee presents a narrower meaning as well, where the natural is that "which is not the material embodiment of deliberate human intention" (Lee 1999, p. 83). The antonym of "natural" thus becomes "artificial," or in Lee's terminology, "artefactual." There are some well-known problems here. First, one can ask whether there is anything left that is "natural" in this sense on our planet. The consequences of humans' actions are vast, and our scope has never been greater than now. Historically, human action has affected the environment more than many people tend to think of, and there does not seem to be any decrease in this. Secondly, the distinctions between man-made things and natural ones are becoming more blurred. Synthetic biology is a case in point.[4] However, I will put these concerns aside here and accept the idea of nature as something unaffected by human actions as meaningful.

Robert Elliot (1982) set the scene for this debate with his essay "Faking Nature." He invites us to consider a case where a mining company removes sand from a beach for the purpose of extracting valuable minerals while offering to restore the area to its original condition afterward, minus the minerals. Elliot argues that even if such restoration efforts were in a sense successful, the full value of the area would not be restored. He uses analogous reasoning from aesthetics to argue his point – that the origin of an object can affect the value of the object. A copy of a Vermeer painting does not possess the value of the original work, even if it were indistinguishable from it.

More explicitly concerned with artefacts is Eric Katz. His thesis is that an artefact, such as a managed forest, "owes its existence to intentions of human agents" (Katz 1993, p. 227). A natural entity is an entity which is not human-created. This distinction, Katz argues elsewhere, "is so clear that I am often perplexed by people who claim that there is no distinction" (Katz 2003, p. 140).

Being the products of human intentions, artefacts essentially are mere means for human ends and do not have a telos of their own. Natural entities, on the other hand, even express autonomy. "Complex holistic natural systems and communities also exhibit autonomy, in that they are independent from external design, purpose, and control" (Katz 1993, p. 230). One might argue that this sense of "autonomy" is rather wide and would make it difficult to account for why and how such "autonomy" should be morally relevant.

---

[3] See also Vogel (2002) and references given there. Vogel himself points out his own, deliberate ambiguity in the use of the phrase "the nature of artefacts" in Vogel (2003, p. 150).

[4] For a discussion, see Preston (2008).

4 Degrees of Naturalness

Among the things that can be learned from this debate, however, one thing stands out: The insistence, by a number of commentators, that naturalness and artefacticity (or some similar-sounding term) are not only distinguishable but also that the difference between them can be a matter of degree. Even if examples of purely natural things are scarce, some things are more natural than others. This holds for areas in the environment – a managed forest is less natural than a pristine one, if there indeed are such forests, and a city is more artefactual (to use Keekok Lee's term) than the Grand Canyon.

This gradual difference between nature and artefacts, where one and the same object might display both characteristics, has as we have seen primarily occurred in the context of originally "natural" objects that have somehow been transformed into being more artefactual than they originally were – for instance, where a pristine forest becomes a managed forest which in turn becomes a park. It is less obvious that this holds also for designed engineering artefacts of the kind that are the paradigmatic examples used by analytic philosophers of technology. However, I think the distinction might be relevant there as well. To see that, let me begin by trying to separate some different senses in which an artefact can be more natural than another. I believe that there are at least four ways in which this can be understood. The difference being gradual, it is perhaps best illustrated by a number of pairwise comparisons of artefacts with different degrees of naturalness:

1. The artefact contains some non-artificial constituent part(s). This is the state expressed by the slogan on my pack of morning muesli: "All natural ingredients!" The muesli is indeed man-made in the sense of being processed – no one expects that you could go out and pick muesli in the field. Someone has at least to mix the ingredients, roast them, and so on. But the constituents are natural. Naturalness in this sense might apply both to macroscopic physical structure (such as shape) or, e.g., chemical structure. As an illustration, consider the following examples: A boat built of wood is more natural than a boat built in aramid fiber and epoxy. And a boat built with grown frames is more natural than one built with steamed frames. In the latter example, natural macroscopic structure is preserved more in the grown frames than in the steamed ones. And it would be even more preserved in a log that had been carved hollow – a primitive boat, but a boat nevertheless.
2. The artefact contains artificial constituent parts, but those constituent parts are identical to ones that can be found in nature. Some pharmaceuticals and flavorings are examples of this. A piece of candy containing vanillin (the most important substance giving vanilla its taste) would thus be more natural than a piece of candy containing a different, hypothetical synthetic chemical that produced a taste indistinguishable from that of vanillin. Or a piece of food containing ascorbic acid used as an antioxidant is more natural than a different piece containing a synthetic antioxidant with relevantly similar properties. Arguably, this could also apply to macroscopic physical structure as well, not only to microscopic or chemical structure.

3. The artefact harnesses a natural process and utilizes it for some purpose. This is thus about the artefact's function being more or less natural, rather than its materials or constituent parts. Consider two devices for purifying water: One relies on a synthetic chemical that degrades the pollutants. The other uses naturally occurring bacteria that feed on the pollutants, thus degrading them. Arguably, the latter artefact is more natural than the former.
4. The artefact produces something that is unpredictable. This is however not because it malfunctions but because it has been designed to do so.[5]

One thing should be noted right away. That an artefact is more natural in any of the senses above does not imply that it is in any way low-tech. Consider (2): Producing a material that is identical to one found in nature may require extremely sophisticated methods. For instance, spider silk is a natural material with several potentially very useful properties, such as elasticity, strength, and biocompatibility. At the same time, producing it using artificial means is difficult and has only recently become a possibility (Stark et al. 2007). The same holds for (3) – It might be much more technically complicated to "tame" a natural process than to utilize an artificial one.

While (1) to (3) at least might seem familiar, despite problems of their own, (4) does require some comment. It is derived from Steven Vogel who introduces the idea in a commentary on the work of Eric Katz. Katz, as we have seen, is concerned with restoration of "natural" areas. Vogel's idea here is that of putting nature – or "wildness," so to speak – back into artefacts. Vogel argues that this might include the purposeful allowing of unpredictable things to occur (Vogel 2003, p. 157). Artefacts in which this might happen are some works of art and gardens, or the "restored" natural areas that Katz discusses. Thus, a garden, like my mother-in-law's, where a part of it deliberately has been left to whatever plants happen to find their way there, would be more natural than a more thoroughly planned one, where every space is planted with designer plants.

There are of course risks involved even with designed unpredictability. Surprises are not always pleasant. In my mother-in-law's garden, there is of course a possibility that the unmanaged part will house some particularly invasive species of weed that threatens to spread to the rest of the garden. On the other hand, a rare butterfly or a particularly beautiful orchid might appear there unexpectedly. It is perhaps not easy to find an example of where this applies to paradigmatic engineered artefacts. Works of art might be more to the point, as Vogel notes.

Nevertheless, from a normative point of view, this sense of naturalness is interesting. Because when naturalness is discussed in relation to artefacts, it is usually taken to be valuable for the very opposite reason. It is technology that is thought to involve (unpleasant) surprises, not nature. The phrase "all natural ingredients" is supposed to connote safety, health, and environmental soundness. It is not supposed to lead me to believe that my morning muesli is likely to behave unpredictably or wildly – quite the contrary, in fact.

---

[5] Malfunctioning is itself a difficult notion, which bears upon the understanding of the nature of artefacts, including the issue of how to individuate them. I will not go into this discussion here, however. See, e.g., Franssen (2006, p. 46f).

5 Should "More Natural" Artefacts Be Preferred to "Less Natural" Ones?

In the late 1800s, Mill wrote that "the word unnatural has not ceased to be one of the most vituperative epithets in the language" (Mill 1988 [1874], p. 11). This is still to some extent true. Condemning all sorts of things – from genetically modified organisms to various sexual activities – as "unnatural" is still common, as well as praising the natural.

Are there any reasons, then, to prefer more natural artefacts to less natural ones? As Chapman (2005) has noted, the idea of the unnatural as somehow wrong has widespread popular appeal, even if it is usually dismissed by philosophers, if mentioned at all.[6]

If there are reasons to prefer more natural artefacts to less natural ones, however, they may be of different kinds. First, there might be what can be termed pragmatic reasons. They need not have anything to do with ethics, of course. It might just be, for instance, that natural artefacts typically are more beautiful than non-natural ones, or the other way around (aesthetic reasons). Or perhaps, the natural artefacts are better qua artefacts than less natural ones. Some designers apparently reason in this way and use nature both in the design process and the marketing of the product. For instance, London-based architect and designer II Hoon Roh writes on his web site: "Structures and forms found in nature, such as beehives and human cells, are models of efficiency and elegance, fashioned over time through the evolutionary process to best serve its purpose … Based on this idea, my primary focus is the Self Form Finding Technique, which is a design method that mimics the building methods already found in nature."[7] I do not have an opinion whether these claims are true (in general), however, and I will not pursue them further.

6 Pragmatic Reasons

One area where a form of naturalness argument has been proposed concerns the management of risks, in particular technological risks. Here, the idea is that natural risks are somehow preferable to technological ones.[8] Technology, in particular technology that involves "tampering" with nature, is something that we need to assess. Perhaps we need to apply one or other version of the precautionary principle (Cf. also Giddens 1999).

---

[6] There are, however, a number of authors who take the issue seriously. For a recent discussion, with relevant references, see Sheehan (2009). Cf. also Takala (2003). We will return to Chapman's own version of the naturalness argument below.

[7] http://www.ilhoon.com/main.html (accessed February 8th 2011).

[8] Hansson (2003). For a discussion of a version of the naturalness argument in the context of risk management, the so-called natural-occurrence view of the *de minimis* principle, see Sandin (2005).

There are good reasons to believe that the common idea that natural risks are less dangerous than technological risk is not justifiable. Again, I am assuming that the distinction is meaningful, which is by no means certain. Is a volcanic eruption a natural risk, when the consequences of the risk are dependent on the fact that humans, using technology, have built their homes on the slopes of the volcano, and so on?

This point is argued by Hansson (2003). Hansson, however, does not consider whether the naturalness of a risk per se is valuable, but only whether it co-varies with a limited number of other relevant properties of the risk. The properties discussed are those of reducibility, adaptation, and non-additionality. Reducibility concerns are based on the idea that natural risks are unavoidable or at least less easily reduced than technological risks and therefore should be given lower priority. According to the argument from adaptation, humans are better adapted to natural risks (e.g., through evolution). The non-additionality argument states that while technological risks add new risks, this is not the case for natural risks.

Hansson dismisses all three arguments. It is quite conceivable, however, that a natural risk is preferable to a non-natural one for other reasons than the ones discusses by Hansson. That is, even if a natural risk is not less dangerous than a technological one, the natural risk might be to be preferred nevertheless. If one believes that natural things possess intrinsic value, for instance, one could argue that a natural risk is to be preferred to an equally large technological risk not because the natural risk is less dangerous but because it is – natural. Phrased in the terms of risk management, this would be to say that the benefits of the risk-taking activity would be increased if the risk is natural.

With that, let us move away from risks to technology. A recent writer who focuses on the naturalness argument is Anne Chapman (2005). Chapman is concerned with technologies rather than with (individual) artefacts, but she admits that naturalness is a matter of degree. She does not, however, provide any clear criteria for determining whether something is natural or not but merely presents a number of "strands to the natural/unnatural distinctions made with regard to human technologies" (ibid.). Those are that (1) natural methods rely on biological processes rather than chemical or physical ones; (2) natural methods are developed through experience from practice while unnatural ones are based on science; (3) we perceive that unnatural methods give us more control over the outcome than natural methods do; (4) unnatural methods, as opposed to natural ones, "aim to give human behavior a freer rein," thus freeing human activity from the constraints nature imposes; and (5) unnatural technologies allow us to create novel things – new kinds of things that never existed in "living nature" (ibid.).

As said, these distinctions are by no means clear, but let us set that aside for the moment. However, on the basis of distinguishing between unnatural and natural technologies, Chapman gives an account of why unnatural technologies are wrong. First, they embody an undesirable attitude to nature. If we use more unnatural technologies, "we lose the encounter with the otherness of nature, diminishing our relationship with it and thus with ourselves" (Chapman 2005). This argument, Chapman admits, is relative. She notes that "there is a vast difference between the primitive 'gardening' of hunter/gatherers … and the attempt to genetically engineer a plant so that it has a particular defined set of characteristics that we want it to have" (ibid.). This is of course entirely correct. But it is very questionable whether the comparison is relevant. Very few people, I think, would seriously consider returning the world to a hunter/gatherer society. And conventional plant breeding (i.e., without GM) has for a long time involved not merely selection of plants with "naturally" existing traits but crossbreeding and the active induction of mutations with the aid of radiation or mutagenic chemicals; it is thus much more invasive than one might believe. The step from this to genetic modification or other biotechnological methods is arguably smaller than previous steps taken in the history of agriculture – such as domestication.

Second, Chapman argues that unnatural technologies, such as GM, involve novelty, i.e., the production of new kinds of things, not only new instances of an existing kind of thing. A newborn mouse in the field is merely new, while a transgenic newborn mouse brought about through IVF is novel, at least the first time a mouse of that particular type is produced. It turns out from Chapman's discussion that it is not in fact novelty that is the real problem, but uncertainty. "Unnaturalness is thus a way of talking about radical uncertainty, or ignorance about the effects of a new technology" (Cf. Hansson 1996). This calls for a precautionary approach, according to Chapman. I do not disagree with Chapman on this, and I have elsewhere discussed and defended precautionary approaches (e.g., Sandin 2006; Sandin et al. 2002). I will not repeat this discussion. But what can be called into question is to what extent this has anything to do with unnaturalness. Chapman argues from unnaturalness to uncertainty to undesirability, and this chain of argument is weak.

Third, she argues briefly that unnatural technologies have detrimental social consequences. For instance, power will be taken away from individual farmers and transferred to agrobiotech companies. As Chapman recognizes, this argument is well known from the debate on GM crops. It is connected with, for instance, the works of Vandana Shiva. Let us accept that this changed power structure is socially detrimental. But Chapman's argument is problematic nevertheless. She argues in the following way:

"However, I do think this is a general feature of unnatural technologies. It arises because natural technologies are generally based on more widely dispersed resources and knowledge and are more amenable to being carried out on a small scale. They are therefore inherently more likely to create a more equal distribution of wealth and power than unnatural technologies" (Chapman 2005).

There is a slide in this passage. Chapman moves from (1) natural technologies generally (which I take to mean "typically," since it does not hold in every case) being based on more available resources, via (2) the detrimental social consequences being a general feature – in possibly a different sense of generality – of unnatural technologies, to (3) natural technologies being inherently likely to create desirable social consequences in the form of increased equality. The problem is again that the connection between a technology's being natural and its having some desirable characteristic is contingent. The same, of course, holds for the connection between unnaturalness and undesirable characteristics. It might well be true that there is a relation between different types of technologies and social consequences of certain kinds. But Chapman does not convincingly show that this is the case for unnatural technologies.

7 Intrinsic Reasons

So, if naturalness in artefact is valuable, it might be so either because naturalness is intrinsically valuable and this value is inherited by the artefact (being partly natural) or because naturalness somehow co-varies with other desirable properties. The proposed arguments for the latter view are, as we saw, rather weak.

But what about the value of artefacts themselves? As noted in the Introduction, it is obvious that artefacts can have instrumental value. A good hammer is a hammer which does the job that hammers are supposed to do, and some hammers do the job better than others. The instrumental value of my new car is greater than that of my old car – it performs the same transport work more reliably, consumes less petrol, and provides a more comfortable ride.

Can then artefacts themselves have intrinsic value, in addition to their instrumental value? I believe so, and I will use a modification of an argument presented by Keekok Lee to argue this point. Lee explicitly denies that artefacts have intrinsic value. She attempts to show this through adapting Richard Sylvan's "Last Man" argument for her purposes.[9] In that argument, Sylvan asks us to imagine a situation in which there is only one person remaining in the world after some catastrophe. Suppose that the last person somehow is able to destroy all other living things, both animals and plants. According to anthropocentrism (or what Sylvan calls "human chauvinism"), the last person does nothing wrong. Nevertheless, many of us have an intuition that the last person acts wrongly. Hence, anthropocentrism is untenable, and nature possesses intrinsic value.

Lee's version of this argument is as follows:

Imagine that you are the last surviving person on Earth (and no equivalent type of consciousness would ever appear in the future). Just before you die, you have the technology to destroy all the non-biotic artefacts in existence. Ought you to do so? Is there any moral or rational constraint upon you to desist? The answer to these questions appear to be no, for the simple reason [...] that artefacts have meaning and significance alone for humans, who are their creators. (Lee 1999, p. 96)

We might begin by noting that while it might be true that artefacts have "meaning and significance" only for humans, there are numerous examples of artefacts that support non-human life. For instance, a sunken ship becomes a "reef" on which non-artificial marine life-forms thrive. If the wreck were to disappear immediately,

[9] Sylvan ([1973] 2009). Richard Sylvan was known as Richard Routley when the original paper appeared in 1973. He has published under both names.

those life-forms would suffer. Not to mention the carpet beetles feasting on my sweaters. So there are instances of artefacts having instrumental value which might not be obvious at the first glance.

But let us set that aside and instead return to the passage quoted above. As we see, on Lee's view, it is not the case that you ought to destroy the non-biotic artefacts,[10] and it is not the case that you ought not to. The actions are completely normatively neutral – both morally and prudentially. Biotic artefacts are different, Lee argues, if somewhat hesitatingly. The reason is that they "appear to possess a residual telos which is independent of that imposed by their human creators" (Lee 1999, p. 96). And if the biotic artefacts are able to maintain themselves, renew, and reproduce themselves, the last person's act of pressing the "destroy" button would no longer be morally neutral.

In addition, she notes that without humans maintaining them, artefacts "would just simply decay and be transformed by nonhuman agents into something else" (ibid.). Thus, "destroying them instantly [...] makes no relevant difference to the outcome." Two things might be noted here:

First, Lee's point that non-human agents transform artefacts indicates that the process of the artefacts losing their artefacthood, if you like, is gradual. At the same time, she seems to want to say that there are no artefacts at all when there are no humans left, since they have meaning only for humans. Artefacts simply cease to exist as artefacts when humans or perhaps some similar type of conscious beings cease to exist, even though the physical remains of the artefacts still remain. This would mean that the artefacts would cease to be artefacts at the instant the last individual that could bestow meaning upon them disappears. Second, there will be considerable differences between how long time it takes for an artefact to decay when there is no maintenance staff around anymore. Some artefacts will decay rather quickly, other artefacts can continue to perform their intended tasks if not indefinitely so at least for a very long period of time.

To me, it is not at all obvious that Lee's adaptation of the Last Man argument succeeds. For consider now instead the following version of the argument: Imagine two scenarios, exactly like the Last Man case. In one scenario, person A goes ahead and presses the "destroy" button. In the other scenario, however, person B – who is like A in all relevant aspects – refrains from doing so. (We also assume that there are no relevant differences in side effects, so that one action hugely benefits or harms other sentient beings, for instance.) When we compare them, is it obvious that A's action is of exactly the same moral worth as that of B? If we think, as I am inclined to, that B's action is better than A's, then Lee's argument does not come through. A's pressing the button might be understandable, but I fail to see that there is no difference at all between A's and B's actions. On my view there is a difference between the actions, such that B's action is better than A's. Those who share this intuition should be led to infer that artefacts can have non-instrumental or intrinsic value, just as those who share the intuition that Routley's Last Person acts wrongly should conclude that anthropocentrism is untenable.

---

[10] For an analysis of living artefacts, see also Siipi (2003).

It is possible that the reason for thinking that there is a morally relevant difference between the actions of A and B in the example is that one infers a difference in character between the agents A and B. Thus, it might seem as if B, in refraining from pushing the button, displays one or other stoic virtue, while A shows a defective character trait, a vice, perhaps some form of rashness. But this, I believe, would be misleading. As the case is presented, we do not know anything about A's and B's reasons for acting. Nor do we know anything else that could enable us to judge their character – the account is simply too thin.

Finally, even if the argument goes through, I do not wish to claim that this establishes that artefacts are morally considerable, merely that they are intrinsically valuable (Cf. Varner 1985). Someone might want to argue that artefacts are morally considerable – there is even a discussion about whether artefacts can be moral agents, entities to which at least partial responsibility can be assigned (Peterson and Spahn 2011. That discussion will, however, have to be postponed to another occasion.

8 Conclusion

In this chapter I have attempted to show how the idea of naturalness as discussed by a number of environmental ethicists might be relevant in the discussion and evaluation also of technical artefacts.

First, the view of "naturalness" as a property that comes in degrees appears tenable for typical engineering artefacts as well, not only for the environmental ethicists' preferred examples of "restored" natural areas or managed forests. Artefacts can be more or less natural in at least four different senses: (1) to what extent the artefact contains natural constituent parts, (2) to what extent it contains constituent parts that are identical to natural ones, (3) to what extent the artefact harnesses natural processes, and finally (4) to what extent the artefact generates intentional unpredictability (or "wildness"). I have argued that a number of pairwise comparisons of artefacts of varying degrees of naturalness can illustrate this. Furthermore, an artefact displaying a higher degree of naturalness does not mean that it is in any way "low-tech" – in fact, it might well be the other way around.

Second, the discussion of whether artefacts have intrinsic value should not be taken as lightly as Keekok Lee does. I have argued that her adaptation of Sylvan's Last Man argument to that effect is unsuccessful and that a version of that thought experiment in fact indicates that artefacts can have intrinsic value. This indicates that the relationship between the value of natural objects and the value of artefacts is more complex than it might appear at first glance.

The discussion about artefacts, norms, and normativity is still in its infancy (Vaesen 2008). It is my firm belief, as I have argued in this present chapter, that there is room for considerable cross-breeding between the field of philosophy of technology and environmental ethics, despite the tradition of the latter field of sometimes being hostile to technology and of the former field of possibly overlooking some of the benefits of the latter.

References

Chapman, Anne. 2005. Genetic engineering: the unnatural argument. *Techné: Research in Philosophy and Technology* 9(2).

Elliot, Robert. 1982. Faking nature. *Inquiry* 25: 81–93.

Franssen, Maarten. 2006. The normativity of artefacts. *Studies in the History and Philosophy of Science* 37: 42–57.

Franssen, Maarten, Lokhorst, Gert-Jan, and van de Poel, Ibo. 2009. Philosophy of technology. *Stanford Encyclopedia of Philosophy*. http://plato.stanford.edu/entries/technology/. Accessed 23 Feb 2011.

Giddens, Anthony. 1999. Risk and responsibility. *Modern Law Review* 62(1): 1–10.

Hansson, Sven Ove. 1996. Decision making under great uncertainty. *Philosophy of the Social Sciences* 26: 369–386.

Hansson, Sven Ove. 2003. Are natural risks less dangerous than technological risks? *Philosophia Naturalis* 40: 43–54.

Katz, Eric. 1993. Artefacts and functions: A note on the value of nature. *Environmental Values* 2: 223–232.

Katz, Eric. 2003. Understanding moral limits in the duality of artefacts and nature: A reply to my critics. *Ethics and the Environment* 7(1): 138–146.

Kroes, Peter, and Anthonie Meijers. 2006. Introduction: The dual nature of technical artefacts. *Studies in the History and Philosophy of Science* 37: 1–4.

Lee, Keekok. 1999. *The natural and the artefactual*. Lanham: Lexington Books.

Leopold, Aldo. 1933. The conservation ethic. *Journal of Forestry* 31(6): 634–643.

Leopold, Aldo. [1949] 1987. *A Sand County Almanac and Sketches Here and There*. New York: Oxford University Press.

McShane, Katie. 2007. Anthropocentrism vs. nonanthropocentrism: Why should we care? *Environmental Values* 16: 169–185.

Mill, John Stuart. 1988(1874). *Three essays on religion: Nature, the utility of religion, theism*. Amherst: Prometheus Books.

Naess, Arne. 1989. *Ecology, community and lifestyle: Outline of an ecosophy*. Translated and revised by David Rothenberg. Cambridge: Cambridge University Press.

Norton, Bryan G. 1996. The constancy of Leopold's Land Ethic. In *Environmental pragmatism*, ed. Andrew Light and Eric Katz, 84–102. London: Routledge.

Peterson, Martin, and Spahn, Andreas. 2011. Can technological artefacts be moral agents? *Science and Engineering Ethics* 17(3): 411–424.

Preston, Christopher J. 2008. Synthetic biology: Drawing a line in Darwin's sand. *Environmental Values* 17(1): 23–39.

Sandin, Per, Martin Peterson, Sven Ove Hansson, Christina Rudén, and André Juthe. 2002. Five charges against the precautionary principle. *Journal of Risk Research* 5(4): 287–299.

Sandin, Per. 2005. Naturalness and de minimis risk. *Environmental Ethics* 27(2): 191–200.

Sandin, Per. 2006. A paradox out of context: Harris and Holm on the precautionary principle. *Cambridge Quarterly of Health Care Ethics* 15(2): 175–183.

Sheehan, Mark. 2009. Making sense of the immorality of unnaturalness. *Cambridge Quarterly of Healthcare Ethics* 18(2): 177–188.

Siipi, Helena. 2003. Artefacts and living artefacts. *Environmental Values* 12: 43–430.

Stark, Margareta, Stefan Grip, Anna Rising, My Hedhammar, Wilhelm Engström, Göran Hjälm, and Jan Johansson. 2007. Macroscopic fibers self-assembled from recombinant miniature spider silk proteins. *Biomacromolecules* 8: 1695–1701.

Strong, David. 1992. The technological subversion of environmental ethics. *Research in Philosophy and Technology* 12: 33–36.

Stueland, Samuel. 1994. The Otis steam excavator. *Technology and Culture* 35(3): 571–574.

Sylvan, Richard. [1973] 2009. Is there a need for a new, an environmental, ethic? Reprinted in *Encyclopedia of environmental ethics and philosophy*, vol. 2, ed. J. Baird Callicott, and Robert Frodeman, 484–489. Detroit: Macmillan Reference USA.
Takala, Tuija. 2003. The (im)morality of (un)naturalness. *Cambridge Quarterly of Healthcare Ethics* 13(4): 15–19.
Vaesen, Krist. 2008. A philosophical essay on artefacts and norms. Diss. Eindhoven: Technical University of Eindhoven.
Varner, Gary E. 1985. The Schopenhauerian challenge in environmental ethics. *Environmental Ethics* 7: 209–229.
Vogel, Steven. 2002. Environmental philosophy after the end of nature. *Environmental Ethics* 24: 23–29.
Vogel, Steven. 2003. The nature of artefacts. *Environmental Ethics* 25: 149–168.
White, Lynn. [1967] 2009. The historical roots of our ecologic crisis. Reprinted in *Encyclopedia of environmental ethics and philosophy*, vol. 2, 440–445, ed. J. Baird Callicott and Robert Frodeman. Gale Cengage Publishing.

Chapter 14
Trust in Technological Systems

Philip J. Nickel

1 Introduction

Technology is a practically indispensible means for satisfying one's basic interests in all central areas of human life including nutrition, habitation, health care, entertainment, transportation, and social interaction.[1] It is impossible for any one person, even a well-trained scientist or engineer, to know enough about how technology works in these different areas to make a calculated choice about whether to rely on the vast majority of the technologies she/he in fact relies upon. Yet, there are substantial risks, uncertainties, and unforeseen practical consequences associated with the use of technological artefacts and systems. The salience of technological failure (both catastrophic and mundane), as well as technology's sometimes unforeseeable influence on our behavior, makes it relevant to wonder whether we are really justified as individuals in our practical reliance on technology. Of course, even if we are not justified, we might nonetheless continue in our technological reliance, since the alternatives might not be attractive or feasible. In this chapter I argue that a conception of trust in technological artefacts and systems is plausible and helps us understand what is at stake philosophically in our reliance on technology. Such an account also helps us understand the relationship between trust and technological risk and the ethical obligations of those who design, manufacture, and deploy technological artefacts.

First, a terminological remark. I will use the terms artefact and technological system in a nonstandard way in this chapter. By an artefact, I mean a relatively discrete object or component part of an object that has been created intentionally for

[1] We may distinguish between welfare interests and ulterior interests. Welfare interests are those interests which are assumed to be had by any person, such as freedom of movement, ownership of property, life, and health. Ulterior interests are related the specific goals of individuals, such as the desire to own a motorboat, to travel to Hawaii, or to start an organic farm. See Feinberg (1984).

P.J. Nickel (✉)
Eindhoven University of Technology, Eindhoven, the Netherlands
e-mail: p.j.nickle@tue.nl

M.J. de Vries et al. (eds.), *Norms in Technology*, Philosophy of Engineering and Technology 9, DOI 10.1007/978-94-007-5243-6_14,

a particular function. By a technological system, I mean a constellation of artefacts that work more or less in conjunction with one another and that create or enable some form of activity. An example of an artefact is a washing machine, or a heating element in a washing machine. An example of a technological system is an array of appliances and component parts, appliance repair equipment, electrical and water supply lines, drainpipes, electrical outlets, detergents, clothing instruction tags, and so forth that together make our activity of washing clothes possible. This contrasts with the standard engineering use of the term system as a set of parts working together for a given function. In order to emphasize my nonstandard use, I will use technological system instead of system by itself.

I begin, in this and the next section, with an explanation and justification of the very idea of trust in technology. It is not an idea widely accepted among mainstream philosophers of trust, despite recent interest in the topic among theorists of technology.[2] Philosophers often distinguish between judgments of reliability on the one hand and genuine trust on the other, pointing out essential differences between these two attitudes.[3] Whereas a judgment of reliability consists of a purely predictive expectation of performance, trust consists of a normative or moral expectation of performance, perhaps associated with characteristic moral "reactive attitudes" such as betrayal, blame, and resentment. Technological artefacts are mentioned as paradigmatic examples of things about which we make judgments of reliability rather than things we can genuinely trust: "In cases where we trust and are let down, we do not just feel disappointed, as we would if a machine let us down. We feel betrayed" (Holton 1994, 66). Another prominent philosopher of trust writes: "Trusting is not an attitude that we can adopt toward machinery. I can rely on my computer not to destroy important documents or on my old car to get me from A to B, but my old car is reliable rather than trustworthy. One can only trust things that have wills…" (Jones 1996, 14). On this view, although we sometimes use the word "trust" to describe our reliance on technology, this is just a relaxed manner of speaking. Genuine trust is an evaluatively rich interpersonal attitude concerning the motives of the trusted person, but reliance is different. It is as if in one's reliance on technology, one couples a judgment of the means available for achieving one's ends with the engineers' definition of reliability: "The probability that an item will perform a required function without failure under stated conditions for a stated period of time" (O'Connor et al. 2002, 2). It seems that the burden of proof falls on one who wants to show that there is an interesting sense in which we can trust technological artefacts going beyond this conception (see Nickel et al. 2010).

---

[2] Recent special issues of *Knowledge, Technology and Policy* and *Ethics and Information Technology* are devoted to trust in technology (Taddeo 2010).

[3] People often contrast trust with mere reliance, but this is not a suitable comparison. Reliance is way of acting, whereas trust is an attitude. For our comparison we must therefore find an attitude associated with "mere" acts of reliance—here I call this a *judgment of reliability*.

There is one obvious strategy available for showing that trust in technology is something more than a mere reliability judgment. It has sometimes been suggested that genuine trust in an artefact is possible so long as one thinks of it as having as its implicit object the people and institutions responsible for the creation and maintenance of the artefact. For example, when I trust a bridge not to crumble as I cross it, the real object of my trust is those who built the bridge and those who are administratively responsible for its maintenance (Origgi 2008, 12). In this way, we do not have to countenance artefacts as free-standing objects of trust, because we can explain trust in artefacts as a kind of trust in persons. Trust in a bridge reduces to or is fully explained in terms of trust in those who are responsible for creating and maintaining the bridge. My trust that my Honda will start is to be explained in terms of my implicit attitude toward the Honda Motor Company and its employees.

This strategy cannot fully explain the idea of trust in artefacts, however. For when I turn the ignition of my car, I rely on the firing of the ignition system, not the creation of a car that has an ignition system that fires. The firing of the ignition system is not an action performed by the employees of the Honda Motor Company nor would we normally say that these employees caused my car to start. It is instead an event caused by the artefact they created, and it is this event that I rely upon. The following principle seems evident:

> (T→R) If I trust some entity E to perform in a particular way, then it must be a salient possibility that I (or somebody I care about) could rely on E to perform in that way.

Since I do not rely on the employees of the Honda Motor Company to cause the ignition system to fire, it follows that I do not trust them to do so. The entity I trust for this performance must be the car itself. To understand trust in technology, we must therefore look at the question of whether I can trust an artefact itself.

I will argue for an account of trust, the Entitlement Account, that makes sense of direct trust in artefacts. The basic idea is that in trusting, we exhibit a willingness to rely for something we value on the performance ϕ of some entity (such as a person, artefact, or system) about which we are committed to saying both that (a) it is worth relying on ϕ and (b) we are entitled to rely on ϕ. The attitude of entitlement is a normative expectation about the performance of the entity, to be explicated in the next section.

This account explicates a core attitude common to several kinds of trust including institutional trust, self-trust (both mental and bodily), and trust in artefacts. Although applying this account in detail to self-trust and institutional trust would lead us too far away from the purpose of this chapter, a brief example is useful to indicate the broader theme. Consider how I trust my memory. Normally, when I make plans today, I trust my own future memory faculties to guide my planned action. If my memory fails, I am likely to feel frustrated or disappointed in a way that reveals an attitude of entitlement to the performance of my own faculties. This feeling of frustration goes beyond what could be explained by a mere judgment of reliability (summed up in the idea that it is worth it to rely on my memory), although this is also a component of self-trust. It exhibits a normative dimension. In the next section, I will develop an argument along these lines concerning trust in technology.

2 Technological Trust and Entitlement to Performance

The first element of trust in a technology T is that T is likely enough to perform in some desired way ϕ, compared with other alternatives, that it is worth staking something of value on ϕ. Without some such attitude, it is hard to explain why a person is disposed to act in a way that presupposes or counts on the performance. But this first element by itself is not sufficient for trust, for in trust one also has a normative attitude toward the entity trusted. Thus, we need a second element in any account of trust in technology, an attitude of entitlement to the technology's performance. In this section I wish to make it plausible that such an attitude exists and that we can make some sense of it folk psychologically. I do not aim here to show that it is legitimate in the deepest sense. Plausible attitude ascription does not depend upon the ultimate legitimacy of those attitudes. For example, philosophers who argue that attitudes of blame are illegitimate on the ground that there is no such thing as freedom of the will, and hence that there is no suitable object of blame, do not thereby infer that blame does not exist.

To show that an attitude exists, the central step is to show that it explains a cluster of phenomenological and psychological observations by providing a rationale for those observations and can be given a plausible description in itself. In the case of our reliance on technology, there are some ordinary phenomena that cannot easily be explained without supposing the existence of an artefact-directed entitlement attitude. This becomes apparent when we focus on negative attitudes toward technological failure such as anger, disappointment, and frustration. All of these negative attitudes can take an artefact or technological system as their object. An academic survey article on anger's role in human development states that it "is associated with infants' attempts to master the physical environment, and … elicits behavioral strategies from infants that serve regulatory functions and contribute to problem solving" (Lemerise and Dodge 2000, 596). Technological artefacts and systems form a large part of the physical environment throughout human development. This is why a car that won't start or a computer that won't print the last chapter of one's dissertation can elicit anger.[4] Frustration and disappointment at the artefact are also common (as indicated by Holton in the passage quoted above, op cit.). These attitudes are distinct from (though compatible with) disappointment in an outcome, anger at oneself, and anger at those who created an artefact.

What is the rationale for such attitudes? If the only attitude we ever took toward an artefact or technological system were that it is sufficiently likely to perform in a certain way that it makes sense to rely upon that performance, there would be no rationale for artefact-directed disappointment or anger. For first of all, it is logically consistent with this judgment that the artefact or system does not in fact perform. The judgment at the core of my original attitude has not been straightforwardly contradicted by later events, since likelihood and nonoccurrence are not strictly

[4] Or even aggression: destruction of objects is a clinical criterion for hostile aggression, although this is often a surrogate for interpersonal aggression (Ramírez and Andreu 2006).

opposed. At most, if an event judged likely does not occur, this requires some revision of one's later judgments of likelihood and an adjustment of one's plans. In itself, it seems implausible that a "cold" epistemological attitude such as prediction would generate object-directed disappointment or even anger when it failed. If anything, its failure should generate other epistemic or intellectual attitudes such as curiosity or puzzlement. Therefore, in order to explain artefact-directed anger and disappointment, it is useful to suppose that people sometimes have a richer normative attitude toward artefacts when they rely on them. I propose an attitude of entitlement, with the artefact's performance as its object.

This key notion, entitlement, has two established philosophical meanings. In one sense, it refers to the positive normative status of an attitudinal state, such that one cannot, other things equal, be criticized for having that attitudinal state. For example, some epistemologists think we are entitled to the belief that physical objects exist, even though they also hold that we have no ultimate justification for this belief since there are no sound arguments conclusively establishing its truth. We cannot be criticized for holding the belief, for it provides support to many other beliefs that are themselves confirmed by experience, consistent with further beliefs we possess, and useful or even indispensible to our lives (Wright 2004).

Second, entitlement is sometimes understood as a right to goods or services from another person. For example, I might be entitled to three apples belonging to another person in virtue of having won a bet with that person. In this sense an entitlement is a Hohfeldian "claim right," consisting of a liberty to receive these apples together with an obligation the other person has to provide me with them (Wenar 2008). In the case where I come across apples not belonging to anybody, my claim right to pick them is a negative right: it consists of a liberty to take them, together with an obligation that others not interfere with my doing so.

It is sometimes assumed that the idea of an entitlement to an artefact's performance must fall into either the epistemic type or the rights type: entitlement is either the warrant for a belief about what I can reasonably predict of the artefact or it is a claim right implicitly directed toward a person or persons responsible for ensuring that it performs a certain way (Franssen 2009). For example, my entitlement to my car's starting is understood either as an epistemic prediction about the car's starting or as a claim right according to which "I can hold the manufacturer, or, in the case of a secondhand car, the car dealer, to his or her part of a deal we made" (ibid., 941). I wish to deny this. The notion of entitlement I wish to elucidate is identical neither with an epistemic entitlement nor with a claim right, even though it has something in common with each. With the epistemic notion of entitlement, it shares the feature that one counts upon a performance (here both practically and epistemically) the likelihood of which cannot be given a conclusive justification. Although on many given occasions one can question or demand justification for one's reliance on a technological artefact or refuse to use the artefact because such justification cannot be found, in our civilization, it would be impossible to do this in general without foregoing the normal pursuit of one's interests. Wittgenstein (1969) made a case that doubt, as well as demands for justification, can only be pursued in everyday life against a backdrop of confident but unjustified belief. Similarly, in a technological age, doubts about technological reliability can only be pursued in everyday life against a backdrop of normal, confident reliance. This generates a default practical entitlement to reliance, in the absence of specific reasons for doubt.

The notion of entitlement I have in mind also has something in common with a claim right. First of all, it is normative: it supposes that the entity ought to perform a certain way, in an evaluative sense. Such an entitlement could be justified in a number of ways, drawing in the first instance on the fact that the artefact has the function of performing in that way—that is, what the artefact is for, what it has been created (and probably advertised and sold) to do, or what it evolved to do. This entitlement could then be used in some circumstances to justify further normative, moral, and legal claims, although it is not identical with those claims. For example, "Failure to ϕ should be compensated for by the manufacturer," "Failure to ϕ is a sign of bad design," "The designer is blameworthy for failure to ϕ." Some of these further claims might indicate rights claims, although they need not do so.

This notion of normative, default entitlement is common to many kinds of trust, including artefactual trust, self-trust, and institutional trust. Here are some examples.

> A baker prepares dough on Tuesday evening, intending to punch it down and roll it out the following afternoon. The baker takes herself to be entitled on that day to rely on her own memory and practical abilities the next day—to complete the action.
>
> Quoting David Lewis, "In my hometown of Oberlin, Ohio, until recently all local telephone calls were cut off without warning after three minutes. Soon after the practice had begun, a convention grew up among Oberlin residents that when a call was cut off the original caller would call back while the called party waited" (1969, 43). Suppose one's call is cut off. In such a situation, one takes oneself to be entitled to rely on the original caller to call back.[5]
>
> I work on a manuscript for an entire day, and at the end of the day, I hit the "save document" button in my word processor, planning to open it again the next day. I take myself to be entitled to the document's latest version being recorded to the hard drive.

A normal way to express this notion of entitlement is to say that some entity (one's memory, one's neighbor, one's computer) is supposed to do something. It can also be expressed using ought or should, where these words are not necessarily taken to express any claim of moral obligation or requirement. I can say, in a normatively rich sense, that I was supposed to remember something or that I ought to have remembered it, without supposing that it is my moral obligation to remember it.

These are not cases in which a person has a claim right to performance, for this would suppose wrongly that there is somebody with a duty to provide the performance in question or to refrain from interfering with the performance, in each case. This would in turn license a moral emotion such as blame or betrayal if performance failed—a right of moral complaint. Such a moral emotion would not be appropriate in these cases, because there is nobody with such a duty. Furthermore,

[5] This entitlement holds even if one is not oneself the person originally called. For example, suppose my brother was phoned but had to leave suddenly just after the call was cut off, and before the conversation had finished. If I am standing nearby, I am entitled to rely on the original caller's calling again.

although belief states are involved in these cases, the epistemic sense of entitlement is too thin to capture the force of these examples. For, first, the issue is what one can depend on in one's actions, not to what extent one has a particular belief. And second, the normativity associated with epistemic entitlement as I explained it above is too weak to account for the attitudes one has about performance. For in these cases one is licensed to be irritated, angry, or disappointed at some entity if performance fails—in the first case above, at oneself, in the second at the caller, and in the third at the computer program. I am not similarly licensed to be angry or disappointed at some entity in the event that some belief in which I am epistemically entitled merely turns out to be false.[6] Furthermore, I am not licensed to be angry or disappointed at just any object for which I stake something of value on its behaving a certain way. For example, I am not licensed to be angry or frustrated at a coin that comes down tails when I have bet heads.[7] My attitude toward the coin in such an instance is different than it would be at a coin that failed to register as proper currency in a vending machine, for example. It is the latter attitude that I wish to describe as an attitude of entitlement, a kind of normative expectation.

I have said that anger or frustration is a normal response to breakdowns of this type of trust. It is often argued, by contrast, that blame and betrayal are the central negative attitudes associated with broken trust (Faulkner 2007; Hieronymi 2008; Holton 1994). But while betrayal and blame have been well studied in connection with trust, little attention has been paid to normatively laden emotional responses toward people who act incompetently but nonculpably, toward one's own performance failures, toward institutions that fail, and toward artefacts that fail. In cases of others' nonculpable failures of competence, disappointment can be highly appropriate. In cases of one's own failure to perform as expected (forgetting, failing to notice something, failing to execute bodily movements successfully, etc.), self-directed frustration is very common. Institutions that fail without the known culpable failure of any individual can induce appropriate anger. Such attitudes are pervasive. Similar reactions toward artefacts also seem appropriate, and although they could also give rise to feelings of blame and betrayal toward the artefact's designer or manufacturer, they are not identical with those feelings.

Our answer to the question whether technological artefacts can normally be an object of anger, say, partly depends on our ability to make sense of an attitude so directed. Charity encourages us to interpret people's attitudes and behaviors in the way that can be given the most rational explication. This may make it difficult to notice the attitude of entitlement that I have placed at the center of trust in technology, because it is easily conflated with a rights claim and with the interpersonal attitudes that accompany that claim. Moral philosophers have been reluctant to admit the possibility that object-directed negative attitudes not conveying a determinate moral complaint can be appropriate. In part, I think, this is simply because they have not

[6] An exception is the case in which I am told something false by another person. But in that case, I have reason to be angry at the person not just because her claim happened to be false but because she wrongly presented the claim as true.

[7] This point is due to Peter Kroes.

paid sufficient attention to the rich array of such attitudes. In part it may be because prevailing moral theories attempt to vindicate emotions of moral complaint but pay little or no attention to object-directed complaints of other kinds.

3 Trust, Risk, and Technological Systems

Let us define risk as the nontrivial (proper) possibility that a serious negative effect will arise in the future, and let us define safety of some process or product as a threshold of low risk associated with the process or product such that it is objectively reasonable to accept the risk.[8] What is the relationship between the following claims about safety and trust?

(A) We have conclusive reason to believe that a technological system S is safe (subjective certainty that S is safe).
(B) We trust S.

These two claims are often both true for a given technological system but neither directly implies the other. On the one hand, even in the case where we think a system like S is a good idea and are in a position to rely on S, (A) does not imply (B). We may have conclusive reason to believe that S is safe but still not trust S. One reason for this is that safety is not the only thing we care about. We may think there are other technologies that perform better than S, or we may think relying on S is not worthwhile given the benefits it produces.

More interesting is the question whether (B) implies (A). Normally it is true that when I trust a technological system I regard it as safe, that is, objectively reasonable to accept the risks it poses. But the notion of "objectively reasonable," and thus the notion of safety as I have defined it, seems to be based on a situation in which one has more than one good option. (A system does not become safe just because there are few alternatives to it, after all.) Trust, on the other hand, might seem somewhat more adaptable to situations in which there are few options. In cases where I do not have many good options besides S and S falls short of safety, we might still want to say that I choose willingly to rely on S and that I have normative expectations about S's performance and thus that I trust S. We are of two minds about the relation between willingness and good options. In one sense, when I hand my wallet to an armed robber, this is voluntary, since I choose that action over likely injury or death. In another sense, I act involuntarily because I have no other reasonable option.[9]

---

[8] These definitions of risk incorporate value notions such as "serious negative effect" and "reasonable to accept." For an argument that the notions of risk and safety are fundamentally normative, see the chapter by Möller in this volume.

[9] This issue is famously discussed by Aristotle, who already sensed the difficulty of settling it in book III of his *Nicomachean Ethics* (Aristotle 1998). Aristotle himself calls coerced actions voluntary because their immediate cause lies in the person who acts.

This difficulty sometimes arises when a new technological system comes into being that completely replaces an old way of pursuing an activity, leaving no easy way to opt out of reliance. For example, a card-based system for gaining access to public transportation might be introduced, or a nuclear power plant might be located near one's long-time place of residence. Suppose I have my doubts about whether these systems are safe (or secure): I feel they may present significant risks to my privacy or life, respectively. Although the question of whether I feel trust toward these systems may still be in order, my disposition to rely on the systems is difficult to assess. I may have little choice but to rely on the system; therefore, in one sense I am so disposed. But I may remain (somewhat) against doing so, were I to have a fully adequate choice in the matter. In that case, my trust or distrust has no immediate practical point but merely registers my feelings, gesturing indistinctly toward some hypothetical situation in which I might be in a position to choose.

Some contexts like this, in which people rely on pervasive technological systems and where doing so would be difficult to avoid, have raised prominent questions about trust in the academic literature. Examples are electronic voting systems (Pieters 2006), electronic patient dossiers (Barrows and Clayton 1996), large-scale energy projects (Viklund 2003), and GMO foods (Frewer et al. 2003). For example, a recent paper describes trust attitudes toward a new system for e-voting in the United Arab Emirates (Salem 2007). As described in the paper, in the UAE there is no election law, and voting is carried out by an electoral college of several thousand citizens appointed by the rulers' courts. The electoral college is more than four-fifths male. In this context, an electronic voting system has been introduced including an electronic system for the registration of electors and candidates, a system for verifying the identity of electors, and a vote casting system with kiosks, a voting database, an encryption system, and a vote counting and result-presentation system. Considering the question of trust and how to enhance it through "knowledge management," the author observes that "Perhaps, the clearest indicator of voter trust and satisfaction with the system" is that the electoral authorities received no requests for a recount and that there were no recorded incidents of voter fraud (ibid., 9). The question of how citizens who were not allowed to vote felt about this technological system is not addressed.

Unless citizens wish to leave the UAE, the existing voting system and its technological realization is a fact of life for them, completely embedded in the emerging machinery of democracy there. It is unclear what meaning we should assign to trust in this context. If we ask whether people trust the system, are we asking whether they acquiesce in a system that poses uncertain risks, that currently violates international norms of democratic political representation, and that they can hardly avoid? Or whether if they were free to choose this system or another available alternative, they would actually be disposed to choose it and would feel entitled to a certain (safe, secure, fair, privacy-respecting) performance from the system? The former is a practically salient question but has little to do with trust, since an affirmative answer is compatible with complacent or even grudging tolerance and a high perception of risk. The latter is not as practically salient but asks a question more in line with the willingness we associate with trust, one that has greater ethical relevance to the legitimacy of the voting system and one that tracks perceptions of safety. It may only be practically relevant to ask this latter question with regard to those instances of reliance on technological systems where one comes close to having a free choice, such as whether to use the Internet (or a particular website) to find health information (e.g., Vedder 2003). It is usually asked against a backdrop of widespread and largely unquestioned reliance on existing technological systems and the artefacts that comprise them, singling out one element for assessment. If this is how we understand questions about trust in technological systems, a conceptual connection between trust and perception of safety is maintained. Trust in a technological system implies imputing a level of safety to that system that one would be willing to accept given a real choice (assuming that the other condition, the normative expectation of a certain performance, is also met).

4 The Ethics of Trust and Technology

In this final section, I discuss the ethics of trust and technology, an area in which ethics and epistemology are closely linked. On the one hand, the ethics of trust requires an ethics of belief. Trust centrally involves doxastic states regarding expected future events. It is partly an estimation that some future event is likely enough to be worth relying upon. But trust is also practical; it inherently disposes one toward the act of reliance, even if the opportunity to do so is not always present. Because of its practical dimension, the ethics of trust is not a pure ethics of belief (if there is such a thing). A person's decision to rely on a given technological artefact or system is influenced by pragmatic factors, such as what alternatives are available (if any), the time it would take to research these alternatives, the benefits she stands to gain if the artefact or system performs as expected, and so on. These pragmatic factors may also affect what she expects from the technology normatively, what she feels entitled to from the technology. Furthermore, the person who relies on technology is rarely in a perfectly free and epistemologically privileged position. She must rely on technology under conditions of time pressure, insufficient cognitive resources, and without full information.

In my view, a person's epistemological and practical condition is improved when she has adequate, sound justification for her trust, going beyond the mere entitlement discussed in section II. I will call this the justified trust ethic. It is a particular application of a plausible minimal rationalist principle which states that in making important decisions, it is best to do so on the basis of adequate reasons. On my account of what trust in technology is, one's justification for it consists in adequate reasons to believe (a) that the technology is sufficiently likely to achieve a performance ϕ that it is worth relying upon it and (b) that one is in fact entitled to ϕ. Having the justification consists of having access to these reasons, in the sense that one grasps and recognizes the reasons. Having the justification does not require being able to articulate, weigh, or support these reasons with more fundamental argument. That would be too intellectualized a conception of having reasons for trust. Even if the notion of a justification is not hyperintellectualized, however, coming to have a justification for one's trust requires scarce cognitive resources such as attention and short-term memory. For this reason, the justified trust ethic must be qualified: coming to have a justification for one's trust sometimes takes up too much time or demands excessive cognitive resources and must be balanced against other factors.

There are two serious arguments against the justified trust ethic. The first is based on skepticism about whether most people can actually have an adequate justification for their trust. They stress that most people are incapable of rationally processing relevant information about risk, probability, and benefit. For example, people tend to value the elimination of risk more highly than the reduction of risk by an equivalent interval (Tversky and Kahneman 1992; Tversky and Fox 1995). They evaluate risks differently when probabilistically equivalent but notationally different ways of presenting information are used (Carling et al. 2009). This and other psychological research suggest that nonexperts (and probably experts as well, when they are not thinking formally) cannot recognize the reasons that support their judgments of risk. Indeed, it is commonly argued (or just assumed) that trust is an alternative to making serious judgments about risk, an easier heuristic that allows one to take risks without doing the cognitive work of gathering information oneself and evaluating it.

Perhaps for these reasons, those who write about trust and technology often presuppose that the designers, manufacturers, and deployers of technology have only two trust-related tasks: first, to make the technology as reliable as possible at doing the things it is supposed to do, where this is understood to include safety—thus to make it trustworthy—and second, to persuade possible users that it is sufficiently reliable and safe, thus trustworthy, using whatever psychologically effective means are available. I will call this the trustworthiness ethic. In arguing that one should design for trust, it is sometimes unclear whether it is the trustworthiness ethic or the justified trust ethic that predominates, since both have as their goal the elicitation of a user or client's trust in some contexts.[10]

In addition to the argument that justifying trust is too demanding, a second positive argument that might be offered in favor of the trustworthiness ethic, and against the justified trust ethic, is that relying on trustworthy technology makes people better off, whereas having a further justification for that reliance does not add any additional benefit. This point is a variant of what has been called the "Meno problem" (named after Plato's dialogue Meno in which the problem was first formulated) for the value of justification (Plato 1961; Pritchard 2008). The idea is that having a justification for one's true belief does not make one any better off than simply having a true belief. To use Plato's example: one may travel to Larissa if one has a true belief about which road to take; having a justification for that true belief does not get one to Larissa any more effectively.

Elsewhere, I make a case for the justified trust ethic, claiming that in certain contexts it generates an obligation on the part of the designer, manufacturer, or deployer of a technology to provide evidence of trustworthiness to people in a

[10] There is also a more specialized sense of "design for trust." Vermaas et al. discuss a special case in which the item to be designed is itself a trust-facilitating information technology (2010).

position to trust that technology.[11] In the remainder of this chapter, I will not repeat that argument but instead expand upon it, defending the justified trust ethic against the two objections to it described above. Thus, first, I will argue that there are often kinds of evidence available relevant to the components of trust that can plausibly be recognized and grasped by the person in a position to trust, and second, that a justification acquired by the person, comprised of this evidence, has value for her. This clears the way to hold that those designing, making, or deploying technology to be trusted must both provide sound evidence of trustworthiness and must not provide unsound or irrelevant evidence that could mislead the potential user. Of course this conclusion is only intended to apply to technologies that are plausibly trustworthy given certain reasonable aims of the potential user; if the technology is not trustworthy and the designer knows this, then, there is no question of providing sound evidence for its trustworthiness to the user.

Evidence that it is worth relying on a given technology depends for its specific content on the aims, interests, and risk aversity of the user. The specific threshold of functioning and safety that the user demands for justified reliance is particular to her. But there are general kinds of information that can provide a sound basis for trust that do not require a nuanced understanding of probabilities, uncertainty, or statistical reasoning. I focus on two relevant types of evidence: first, evidence that failure to perform will (often) lead to an effective sanction. This makes it clear that it is in the designer's interest to ensure performance. The effective sanction could consist of a punishment, reputational damage, or loss of future opportunities. The evidence needs only to draw attention to the institutional structures that make such sanctions possible, not to the possibility of specific performance failures. The second type of evidence consists of an indication that other parties, independent of the designer and sharing the values of the user, are willing to stake their reputations on the technology's performance. These two types of evidence are sound reasons for trust because they indicate why the designer has a strong interest in serving the interests of the user. They relate both to the likelihood that the technology will perform a certain way, as well as the basis for the user's perceived entitlement to that performance. Russell Hardin describes this as "encapsulation of interests" and thinks of it as the paradigmatic reason for taking the attitude of trust (Hardin 2006). Some social scientific accounts of the reasons of trust, emphasizing reputational and institutional factors, support this view (Coleman 1990; Pettit 1995).

It is highly plausible that in some cases sound information of these types could be made available to the user and that the user could recognize it as such. Neither form of information depends on difficult, psychologically unreliable forms of probabilistic reasoning. For example, the user can be shown that an artefact is certified by a governmental authority with the power to sanction failures of performance. At the same time, the second component of trust in technology, the normative expectation, can provide a focal point for communication about what (level of) performance the user should expect. Instead of leaving this implicit, designers can make it clear to the user that

[11] Nickel 2011. I will simply say "designer" and "user" in what follows for the sake of economy.

certain obvious aspects of performance such as safety and privacy meet established public norms, and it can be made clear to the user what the technology is (and is not) designed to do.

However, so long as a given technology is in fact trustworthy, one might wonder in what way it matters how the designer convinces potential users of this fact. Or to put the point in terms of the user's point of view: so long as she has a true belief that the technology is trustworthy, why is it important that she has access to sound reasons for this belief? One crucial reason is that offered by Plato himself in response to the Meno problem: beliefs held without reason or held on the basis of unsound reasons are brittle, insufficiently "tied down" to ensure a stable attitude. Suppose I decide to join a social networking Internet site, and my reason for trusting the site to keep details about my telephone number, relationship status, and political affiliation private is merely that my friends use it regularly and seem to have no such problems. Suppose later one of my friends reports that her private details were made visible to the general public, in just the way I hope to avoid happening in my own case. Suppose that, unknown to me, this happened because my friend made a culpable mistake, but she tells me it is because of a design flaw in the software. If my only reason for believing that the site will keep my details private is the experience of my friends, then at this point my trust may vanish—but if my trust were (also) based on an awareness that an independent body verifies that the site meets established privacy standards, then I might think twice before abandoning my trust in the site on the basis of my friend's experience. Although this is perhaps a trivial case, we can easily imagine similar cases in which the technology in question was a radiation treatment for cancer, a boat engine, or a new business data management system.

A second reason for providing sound evidence of trustworthiness is that doing so upholds a principle of respect toward users of technology. A widely held ethical principle holds that when important decisions are to be made, those affected by or making these decisions should have the opportunity to give their free, considered consent to the decision, using relevant information. This is known as the principle of respect or the principle of informed consent.[12] Although technology users may be limited in their ability to calculate risk decision theoretically, they have what Gigerenzer and others have defended as "bounded rationality" (Gigerenzer and Selten 2002), allowing them to make rational decisions in many typical contexts. This suggests that, instead of bypassing or ignoring the rational capacities of technology users, it is better to target information to their abilities. Furthermore, some users of advanced technological systems or stakeholders in their use have a

[12] The principle of informed consent is best known from the domains of research ethics and medical ethics (National Commission 1979). My suggestion here differs from the usual application of the principle of informed consent in three ways: first, I am not suggesting that it be thought of primarily as an institutional or legal norm; second, I am applying it to any technological artefact or system that has significant possible impacts for the user, where the user is in a position to choose that technology; and third, I am applying it to the reasons for trust, rather than to the risks and benefits associated with the technology itself. (In the usual application of the principle, the technology in question would be a research protocol or a medical therapy.)

better than average ability to make rational judgments about technical reliability and have more knowledge than average about reasonable norms of performance that apply to a given technology. They deserve to have a reasonable opportunity to form a well-grounded attitude of trust or distrust supplemented by evidence about the interests of the technology's designers.

In this chapter I have argued for a distinctive philosophical conception of trust in artefacts and technological systems. I have also argued that this conception invites a new understanding of our way of relying on the designed environment under conditions of risk and that it allows us to make a distinctive ethical point about technology design. Much more work needs to be done exploring this topic, including some work that is empirical or interdisciplinary in nature. For example, it would be good to have a better grasp of the psychological basis on which people develop normative expectations of artefact performance and on how communication affects these expectations. The framework presented here is an a priori starting point for such investigations.

References

Aristotle. 1998. *The Nicomachean Ethics*. Oxford: Oxford World's Classics.

Barrows, Jr., Randolph, C., and Paul D. Clayton. 1996. Privacy, confidentiality, and electronic medical records. *Journal of the American Medical Informatics Association* 3(March/April): 139–148.

Carling, Cheryl L.L., et al. 2009. The effect of alternative summary statistics for communicating risk reduction on decisions about taking statins: A randomized trial. *PLoS Medicine* 6(8): 1–10.

Coleman, James. 1990. *Foundations of social theory*. Cambridge, MA: Harvard University Press.

Faulkner, Paul. 2007. On telling and trusting. *Mind* 116: 875–902.

Feinberg, Joel. 1984. *The moral limits of the criminal law, vol. 1: Harm to others*. Oxford: Oxford University Press.

Franssen, Maarten. 2009. Artefacts and normativity. In *Handbook of the philosophy of science, vol. 9: Philosophy of technology and engineering sciences*, ed. Anthonie Meijers, 923–952. Amsterdam: Elsevier.

Frewer, Lynn J., Joachim Scholderer, and Lone Bredahl. 2003. Communicating about the risks and benefits of genetically modified foods: The mediating role of trust. *Risk Analysis* 23: 1117–1133.

Gigerenzer, Gerd, and Reinhard Selten. 2002. *Bounded rationality: The adaptive toolbox*. Cambridge, MA: MIT Press.

Hardin, Russell. 2006. *Trust*. Malden: Polity.

Hieronymi, Pamela. 2008. The reasons of trust. *Australasian Journal of Philosophy* 86: 213–236.

Holton, Richard. 1994. Deciding to trust, coming to believe. *Australasian Journal of Philosophy* 72(1): 63–76.

Jones, Karen. 1996. Trust as an affective attitude. *Ethics* 107: 4–25.

Lemerise, Elizabeth A. and Kenneth A. Dodge. 2000. The development of anger and hostile interactions. In *Handbook of emotions*, 2nd edn., ed. Michael Lewis and Jeannette M. Haviland-Jones, 594–606. New York: The Guilford Press.

Lewis, David. 1969. *Convention*. Cambridge, MA: Harvard.

National Commission for the Protection of Human Subjects of Biomedical and Behavioral Research. *The Belmont Report*. Available at the United States National Institutes of Health Website, <http://ohsr.od.nih.gov/guidelines/belmont.html>. Original report published 1979.

Nickel, Philip J. 2011. Ethics in e-Trust and e-Trustworthiness: The case of direct computer-patient interfaces. *Ethics and Information Technology* 13: 355–363.

Nickel, Philip J., Maarten Franssen, and Peter Kroes. 2010. Can we make sense of the notion of trustworthy technology? *Knowledge Technology and Policy* 23: 429–444.

O'Connor, Patrick D.T., David Newton, and Richard Bromley. 2002. *Practical reliability engineering*. New York: Wiley.

Origgi, Gloria. 2008. *Qu'est-ce que la confiance?* Paris: VRIN.

Pettit, Philip. 1995. The cunning of trust. *Philosophy and Public Affairs* 24(3, Summer): 202–225.

Pieters, Wolter. 2006. Acceptance of voting technology: Between confidence and trust. In *iTrust 2006, LNCS 3986*, ed. K. Stølen et al., 283–297. Berlin: Springer-Verlag.

Plato, *Meno*, Trans. W.K.C. Guthrie. 1961. In *he Collected Dialogues of Plato*, ed. Edith Hamilton and Huntington Cairns, 353–384. Princeton: Princeton University Press.

Pritchard Duncan. 2008. The value of knowledge. *The Stanford Encyclopedia of Philosophy* (Fall 2008 Edition), ed. Edward N. Zalta. URL = http://plato.stanford.edu/archives/fall2008/entries/knowledge-value/

Ramírez, J.M., and J.M. Andreu. 2006. Aggression, and some related psychological constructs (anger, hostility, and impulsivity): Some comments from a research project. *Neuroscience and Biobehavioral Reviews* 30: 276–291.

Renn, Ortwin. 2008. *Risk governance: Coping with uncertainty in a complex world*. London: Earthscan.

Salem, Fadi. 2007. Enhancing trust in E-voting through knowledge management: The case of the UAE. In *Managing knowledge to build trust in government*, ed. H. Qian, M. Mimicopoulos, H. Yum. New York; United Nations Department of Economic and Social Affairs (UNDESA).

Taddeo, Mariarosaria. 2010. Trust in technology: A distinctive and a problematic relation. *Knowledge, Technology and Policy* 23: 283–286.

Tversky, Amos, and Craig R. Fox. 1995. Weighing risk and uncertainty. *Psychological Review* 102: 269–283.

Tversky, Amos, and Daniel Kahneman. 1992. Advances in prospect theory: Cumulative representations of uncertainty. *Journal of Risk and Uncertainty* 5: 297–323.

Vedder Anton. 2003. Betrouwbaarheid van internetinformatie. In *Jaarboek ICT en samenleving 2003: De sociale dimensie van technologie*, J. de Haan en J. Steyaert, 113–132. Amsterdam: Boom/SCP.

Vermaas, Pieter E., Yao-Hua Tan, Jeroen van den Hoven, Brigitte Burgemeestre, and Joris Hulstijn. 2010. Designing for trust: A case of value-sensitive design. *Knowledge, Technology and Policy* 23: 491–505.

Viklund, Mattias J. 2003. Trust and risk perception in Western Europe: A cross-national study. *Risk Analysis* 23: 727–738.

Wenar, Leif. 2008. Rights. In *The Stanford encyclopedia of philosophy (Fall 2008 Edition)*. Edward N. Zalta (ed.). URL = <http://plato.stanford.edu/archives/fall2008/entries/rights/>

Wittgenstein, Ludwig. 1969. *On certainty*. New York: Harper.

Wright, Crispin. 2004. On epistemic entitlement. *Proceedings of the Aristotelian Society Supplementary Volume* 78: 167–212.

Index

A
ABET. *See* Accreditation Board for Engineering and Technology (ABET)
Acceptance, 9, 24, 25, 55–64
Accreditation Board for Engineering and Technology (ABET), 85
Action, 2, 21, 38, 59, 68, 83, 111, 121, 138, 155, 175, 193, 211, 225
Applied science, 36–38, 53
Approachability, 9, 84, 94–97
Audi, R., 19, 28, 45, 139, 146
Axiological commensurability, 79–81

B
Belief, 9, 10, 16, 19, 22–29, 36, 42, 45–47, 49, 52, 55–64, 138, 179, 204, 205, 210, 219, 227, 229, 232, 233, 235
Brand (artefact), 164–165
Bratman, M.E., 58, 59, 88, 123
Brundtland Commission, 75
Bunge, M.A., 37, 38, 40, 41, 49, 53

C
Category-specified value statement, 37, 106–108, 115
Chapman, A., 214–217
Classificatory value statement, 104
Cohen, L.J., 37, 57, 58
Comparative value statement, 104
Consequentionalism, 194–196, 206
Consistency, 9, 22, 45, 49, 52, 53, 84, 96–97
Constraints, 49, 51, 75, 85–87, 91, 96, 135, 215, 217
Constructivism, 173
Context-dependent, 58, 59, 62–64, 76
Correspondence principle (CP), 203–206
Customer needs, 83–87, 92–94, 97, 128

D
Degrees of naturalness, 208, 212–214, 219
Degrees of rightness, 193–206
Design knowledge, 9, 17, 19, 21–22, 25, 26, 29
Design science, 37, 40, 41, 50
Dewey, J., 27, 72
Distributive normativity, 182–185

E
Edison, T.A., 15, 16
Efficacy, 3, 28
Efficiency, 3, 28, 44, 48, 53, 70, 71, 75–78, 80, 106, 214
Elliot, R., 211
Engel, P., 58, 59, 64
Entitlement, 10, 61, 225–230, 232, 234
Environmental ethics, 208–210, 219
Epistemic uncertainty, 181–186
Eriksson, B., 199
Ethics of trust, 10, 232–236
Evaluability, 9, 84, 93–94, 97

F
Fairness, 76, 80–81
Flexibility, 75–76
Functionality, 10, 69, 78, 79, 108, 122, 123, 156
Functional knowledge, 9, 17–19, 21, 23, 26, 29, 62, 63, 137–139, 141, 146, 151, 152

M.J. de Vries et al. (eds.), *Norms in Technology*, Philosophy of Engineering and Technology 9, DOI 10.1007/978-94-007-5243-6,

Functional nature, 59, 61–63
Functional objective, 68, 69, 71, 74, 77
Functional subcategory, 114–116, 156–158
Functional value statement, 1107–111
Function ascription, 60, 64, 125, 138, 142, 144, 145

G
Gettier, 19, 23
Goodness, 119, 122, 123, 125, 127, 134, 155–169

H
Hansson, S.O., 5–10, 37, 72, 83, 84, 86, 93, 96, 98, 104, 106, 108, 110, 111, 115, 156, 164, 165, 169, 174, 176, 214–216

I
Impact objective, 68, 69, 71, 74, 77
Installability, 9, 132–133
Instrumental function, 138–148, 150, 152
Instrumental goodness, 119, 122, 125, 162, 163
Instrumental value, 42, 105, 162–165, 167, 207, 217, 218
Interference, 9, 90, 123–125, 127, 129, 133
ISO norms, 122

J
Justified true belief, 9, 19, 22–27, 29, 56

K
Katz, E., 211, 213
Kindhood, 155–169
Kinds (artefact), 10, 158–165, 168
Know-how, 17–20, 22, 25–29, 46, 51
Knowing-how, 46
Knowing-that, 46

L
Last person argument, 208
Laudan, L., 83, 94
Leopold, A., 209, 210
Lockhart, T., 199
Lottery paradox, 204, 205

M
Malfunctioning, 26, 63, 122, 138, 148–150, 155, 161, 213
Manufacturability, 9, 129–132, 134
Marketability, 9, 128–129
McAllister, J.W., 18
Means-ends, 45, 48, 122–124, 139, 140, 142, 144, 145, 147
Mental skills, 26
Mill, J.S., 199, 200, 210, 211, 214
Mitcham, C., 17, 19, 35, 55
Mokyr, J., 49
Morton, A., 19

N
Naess, A., 209, 210
Natural concept, 138, 173, 175, 185, 186, 189
Naturalistic reduction, 174, 176, 181, 183, 184, 186
Nelson, R.R., 37, 50–52
Non-anthropocentric ethics, 208
Normal token, 144–146, 148–150
Normative concept, 38, 71, 111, 175
Nozick, R., 72, 88, 196, 197

O
Objective risk, 4
Operational knowledge, 46, 48, 143
Operational principle, 20, 21, 159, 160

P
Perceived risk, 4
Physical nature, 59–64
Pitt, J.C., 57
Plan analysis, 37, 41–44, 46, 48–53
Polanyi, M., 19
Pollock, J.L., 19, 41, 49
Poorness, 157, 161, 162, 165, 167
Practical rationality, 40, 41, 47–49, 53, 72, 139
Practical reasoning, 9, 10, 47, 57, 58, 137, 138, 145, 151, 152
Precision, 9, 68, 73, 75–77, 79–81, 84, 91–93, 97, 108
Prescriptive knowledge, 9, 17, 19–22, 24–26, 29, 48, 64
Probabilistic concept of risk, 174, 176, 178, 188

Probability, 4–8, 149, 174, 176, 181, 182, 184, 193–196, 198, 204, 205, 224, 233, 234
Procedural knowledge, 17–20, 37, 46–48, 52, 53

Q
Quality, 3, 9, 19, 20, 44, 45, 68, 69, 74, 75, 77, 109, 120–130, 134, 155, 158, 162, 165, 168, 177, 180, 182, 186, 187, 199

R
Rationality, 4, 5, 40–42, 45, 47–51, 53, 68, 71–75, 83, 84, 94, 95, 97, 121, 139, 235
Recoverability, 9, 133–134
Requirements, 3, 8, 20, 35, 40, 48, 49, 52, 55, 68, 77, 79, 83–87, 91–93, 95–97, 127, 129, 133, 160, 162, 166, 168, 228
Risk analysis, 4–6, 173–176, 179, 189–190
Risk assessment, 4–5, 174, 176, 180
Routley, R., 208, 217
Ryle, G., 19, 25

S
Safety, 2, 4–8, 20, 45, 57, 61, 63, 64, 69–71, 75, 77, 78, 80, 85, 94, 95, 97, 112, 120–121, 132, 133, 173–190, 213, 230, 232–235
Searle, J.R., 23, 24
Sloman, A., 199
Socio-technical system, 20, 39
Specifications, 3, 20, 21, 36, 39, 44, 71, 83, 84, 87, 91–94, 97, 105, 106, 108, 112, 114, 121, 130, 135, 141, 143–146, 150, 157, 160–162, 164–166
Standard, 8, 9, 16, 20, 22–27, 29, 39–41, 44–49, 53, 55, 56, 64, 68, 79–81, 86, 87, 103–106, 109, 110, 112, 114, 119–124, 126, 132, 134, 135, 158–160, 163–165, 174, 176, 185, 199–201, 224, 235
Structural knowledge, 17, 19–22
Sustainability, 3, 69, 71, 76, 77, 79, 80, 133
Synoptic values, 105

T
Technological rule, 9, 37–41, 48–50, 53
Technological system, 2, 3, 10, 223–236
Theoretical rationality, 47
Token (artefact), 145, 148, 152
Trade-off, 3, 9, 68, 69, 75–77, 79, 81, 85, 97, 129, 135
Transportability, 9, 132–133
Trust, 10, 61, 223–236
Trustworthiness, 233–235
Tuomela, R., 58
Type (artefact), 59, 60, 92, 108, 141–145, 148

U
Uncertainty, 7, 73, 76, 79, 80, 181–186, 199, 216, 223, 234
Usefulness, 3, 36, 165–167
Use plan, 10, 42–47, 50, 85, 141–144, 150, 165–168

V
Value concepts, 111, 175, 183
Value statements, 5, 37, 103–112, 115, 156
van Aken, J., 40, 41, 51
Viewpoint-specified value, 105, 106
Vincenti, W.G., 19, 39, 51
Vogel, S., 211, 213
von Wright, G.H., 42, 61, 105, 119, 121, 138, 140, 167, 227

W
Williams, B., 186, 188
Williamson, T., 26, 201
Winner, L., 1
Winter, S.G., 37, 50, 51

MIX
Papier aus verantwortungsvollen Quellen
Paper from responsible sources
FSC® C105338

If you have any concerns about our products,
you can contact us on
ProductSafety@springernature.com

In case Publisher is established outside the EU,
the EU authorized representative is:
Springer Nature Customer Service Center GmbH
Europaplatz 3, 69115 Heidelberg, Germany

Printed by Libri Plureos GmbH
in Hamburg, Germany